Bioactive Peptides Produced by Limited Proteolysis

Colloquium Series on Neuropeptides

Editors

Lloyd D. Fricker, Ph.D., Professor, *Department of Molecular Pharmacology, Department of Neuroscience, Albert Einstein College of Medicine, New York*

Lakshmi Devi, Ph.D., Professor, Principal Investigator, Director, Interdisciplinary Training in Drug Abuse, Associate Dean for Academic Enhancement and Mentoring, *Mt. Sinai School of Medicine, New York*

Communication between cells is essential in all multicellular organisms, and even in many unicellular organisms. A variety of molecules are used for cell-cell signaling, including small molecules, proteins, and peptides. The term 'neuropeptide' refers specifically to peptides that function as neurotransmitters, and includes some peptides that also function in the endocrine system as peptide hormones. Neuropeptides represent the largest group of neurotransmitters, with hundreds of biologically active peptides and dozens of neuropeptide receptors known in mammalian systems, and many more peptides and receptors identified in invertebrate systems. In addition, a large number of peptides have been identified but not yet characterized in terms of function. The known functions of neuropeptides include a variety of physiological and behavioral processes such as feeding and body weight regulation, reproduction, anxiety, depression, pain, reward pathways, social behavior, and memory. This series will present the various neuropeptide systems and other aspects of neuropeptides (such as peptide biosynthesis), with individual volumes contributed by experts in the field.

Published titles

(for future titles please see the website, www.morganclaypool.com/page/lifesci)

Bioactive Peptides Produced by Limited Proteolysis
Antonio C. M. Camargo, Beatriz L. Fernandes, Lilian Cruz and Emer S. Ferro
www.morganclaypool.com

ISBN: 9781615043682   paperback

ISBN: 9781615043699   ebook

DOI: 10.4199/C00056ED1V01Y201204NPE002

A Publication in the

COLLOQUIUM SERIES ON NEUROPEPTIDES

Lecture #2

Series Editors: Lloyd D. Fricker, Albert Einstein College of Medicine, and Lakshmi Devi, Mt. Sinai School of Medicine

Series ISSN

ISSN   2166-6628   print
ISSN   2166-6636   electronic

Bioactive Peptides Produced by Limited Proteolysis

Antonio C. M. Camargo
Laboratory of Applied Toxinology, Butantan Institute, São Paulo, 05503-900

Beatriz L. Fernandes
Biotechnology Program, Biomedical Science Institute,
University of São Paulo, 05508-000; SP, Brazil

Lilian Cruz and Emer S. Ferro
Department of Cell Biology and Development, Biomedical Science Institute,
University of São Paulo, 05508-000; SP, Brazil

COLLOQUIUM SERIES ON NEUROPEPTIDES #2

MORGAN & CLAYPOOL LIFE SCIENCES

ABSTRACT

Proteins are considered supremely important for the organization, survival, and functioning of living organisms. They were considered stable and static molecules until the early 1940s, when Rudolph Schoenheimer demonstrated that proteins exist in a constant dynamic process of synthesis and degradation (proteostasis), absolutely essential for life. Since then, general and limited protein degradation became some of the most fascinating aspects of biological sciences. This book is focused on a particular aspect of protein degradation, namely, limited proteolysis, which gives rise to bioactive peptides as a result of the enzymatic action of proteinases and peptidases, which are enzymes that hydrolyze specific peptide bonds of proteins and peptides, respectively. In a broad sense, bioactive peptides are any fragment of endogenous or exogenous proteins able to elicit either physiological or pathological activities. Here, we aim at presenting to the readers that bioactive peptides are not merely produced through random processes during protein degradation, but rather through a well-organized enzymatic process that is deeply integrated in the homeostatic processes of living organisms.

KEYWORDS

limited proteolysis, proteostasis, protein processing, bioactive peptides, immunogenic peptides, proteasome, oligopeptide, oligopeptidases

Acknowledgments

Authors are thankful to Prof. Lloyd D. Fricker, Albert Einstein College of Medicine, Bronx, NY, USA, for critical reading of this manuscript. This work was primarily supported by the Brazilian National Research Council (CNPq; grant 559698/2009-7 - Rede GENOPROT) and partially by the University of São Paulo (Grant# 2011.1.9333.1.3, NAPNA). A.C.M.C., L.C. and E.S.F. are fellowship recipients from CNPq. The views and concepts expressed in this book have been markedly influenced by A. Camargo's post-doctoral training at Brookhaven National Laboratory under the guidance of L.J. Greene (1970–72).

Contents

CHAPTER 1

Overview and Historical Background

"The living world that teems around us, the world of species, individual organisms, organs, tissues, and cells, can be viewed as the manifestation of a vast fluid array of protein molecules, each appearing and disappearing in the proper place at the proper time" [1].

1.1 OVERVIEW

The stability of the genome is crucial for the living organisms. Whenever the DNA molecule is broken or damaged, it has to be healed, but never degraded [2]. In contrast, protein degradation is critical for the survival of microorganisms, plants, and animals. Life manifests itself through a permanent process of global and/or selective protein synthesis and degradation. The lifetimes of proteins differ greatly from each other, depending on their role. Some structural proteins remain unchanged for days or years, whereas exogenous and regulatory proteins survive for only a few minutes [3, 4, 5].

When the DNA double-helix was shown to contain all information required for living beings, the molecular steps leading to protein synthesis became the most glamorous and intensive area in biological research. Scientists focused mostly on how the genetic code is transcribed to RNA and translated to proteins. Surprisingly, few scientists dedicated themselves to the pioneering studies of autolysis, i.e., the degradation of self-proteins into their constituent amino acids. Autolysis was supposed to provide essential amino acids during starvation, whenever the exogenous source failed to provide them from food digestion.

The importance of challenging the old concept of protein stability was pioneered by David Rittenberg and Rudolph Schoenheimer. In the early 1940s, their work demystified the stability of constituent proteins of living organisms. They used a newly developed technological method, consisting of incorporating ^{2}H and ^{15}N isotope tags into nascent proteins and followed the half life of the labeled protein in living organisms [6]. This experiment was sufficient to contradict the belief that somatic proteins were stable and static. Schoenheimer, then at Columbia University in New

York City, presented his revolutionary findings at the prestigious Edward K. Dunham Lecture at Harvard University on "The Dynamic State of Body Constituents" [7].

Since that time, the dynamic state of proteins outside and inside the cells has led us to the modern biomedical sciences. In the 1950s, a major fundamental contribution gave morphological and biochemical support to Schoenheimer's concept: the discovery that lysosomes participate at intracellular proteolysis. This discovery identified the location where a large number of intracellular proteolysis occurs and how lysosomal cathepsins contribute to the reutilization of amino acids for the biosynthesis of new proteins [5, 8].

However, the degradation process is not restricted to the normal protein constituents of the body. It is extended to abnormal, useless, unfolded, or aggregated proteins, in a housekeeping process, necessary to avoid accumulation of deleterious proteins within the cell, which could cause serious diseases, such as Alzheimer's disease [see 3]. The concept of general proteolysis is therefore applied to designate the degradation of food proteins, the regulation of protein half-life, the performance of housekeeping, the delivery of amino acids during starvation, the destruction of aggressor organisms, and finally, the integration of the pathways of catabolic processes with the anabolic reactions, essential programs of protein homeostasis (proteostasis).

The proteolytic enzymes in charge of proteolysis are the proteinases dedicated to protein degradation, and the peptidases, which are restricted to protein fragments or peptides. These two classes of hydrolytic enzymes are responsible for the transformation of proteins into their constituent amino acids. The enzymes participating at general proteolysis are either located in the extracellular or intracellular compartments. The extracellular general proteolysis is performed by oral and gastro-intestinal proteolytic enzymes (pepsin; trypsin; chymotrypsin; carboxypeptidase A1, A2, and B1; and aminopeptidases) and by enzymes responsible for the cardiovascular homeostasis (renin, kallikreins, thrombin, plasmin, angiotensin-converting enzyme, for instance). The protein substrates, which will be considered here, are derived from food, from constituent proteins, or from pathogenic aggressors. Outside of the cells are the proteins from food, which are subjected to digestion by oral and gastrointestinal enzymes or endogenous proteins, such as plasma kininogen and angiotensinogen, subjected to processing by the chain of proteolytic events occurring in blood clotting, complement activation, fibrinolysis, for instance. On the other hand, the intracellular general proteolysis relies, among others, on cathepsins, multicatalytic proteases (proteasomes), caspases, and a number of cytosolic peptidases including aminopeptidases and oligopeptidases. The protein substrates originate from endogenous or exogenous sources and from protein fragments generated by the proteolytic enzymes. The hydrolysis of proteins to their constituent amino acids is the ultimate ending of general proteolysis. However, as far as protein turnover is concerned, the general proteolysis should be connected to protein synthesis in order to keep the physiological concentration of each specific protein at the proper place and at right time.

In the middle of the last century, another fundamental proteolytic process, distinct from general proteolysis, was recognized by Linderstøm-Lang, calling it Limited proteolysis [9]. Limited proteolysis gives rise to functional proteins or inactive and active fragments. Specific and limited proteolytic processes performed by special proteinases and/or peptidases are able to convert inactive proteins or peptides into active ones. Limited proteolysis is one of the key processes of post-translational modifications of proteins, thereby participating in the physiological and pathological processes of life.

This book deals with a particular aspect of limited proteolysis that gives rise to bioactive peptides, which are defined as protein fragments that elicit biological responses. The review of the literature on the bioactive peptides prompted us to delineate common features, regardless of the diversity of proteolytic processing, which results in peptide structures able to be recognized by specific regions of macromolecules or macromolecule complexes, thereby provoking metabolic and morphological changes and immune defense. This approach led us to hypothesize that the naturally occurring proteins are potential precursors of bioactive peptides, once they are subjected to the appropriate limited proteolysis process.

Much evidence supports the hypothesis that peptides derived from proteolysis are messengers of complex biological processes. This particular view of limited proteolysis would greatly broaden the genetic program endowed to each protein, thus setting the protein-fragments not simply as one step before the end of the biological function of proteins, but rather as the beginning of a yet poorly understood side of biology, that is, the dynamics of the network of processes that lead to cellular homeostasis.

Very exciting new results allowed us to speculate, whether limited proteolysis is intrinsically connected to general proteolysis. We shall present some studies showing that the bioactive products of limited proteolysis are important players in regulating the dynamic state of proteins in living organisms [see Chapter 3; 10].

1.2   HISTORICAL BACKGROUND

The chemical investigation of protein and peptide structure forcibly had to deal with proteolytic enzymes, both as tools to examine the protein and peptide constituents, and as means to study mechanisms of catalysis. Joseph S. Fruton, in 1938 [11] recognized that the fundamental step had been given by Waldschmidt-Leitz, who evidenced Emil Fischer's proposal that the chemical action of proteinases was to catalyze the hydrolysis of the peptide bond (–CO–NH–) in proteins. Besides demonstrating that the amino acids were the basic constituents of proteins, the proteolytic enzymes explained how these digestive hydrolytic enzymes provided amino acids for the *de novo* synthesis of proteins. In his article, Fruton also acknowledged that the digestion of proteins required two distinct classes of proteolytic enzymes in order to release free amino acids: the proteinases, which are able to split proteins into fragments (i.e., pepsin, trypsin, cathepsins), and the peptidases, which hydrolyze peptides, but not proteins. This nomenclature was adopted, along with the definitions, throughout this book.

Other proteolytic processes, however, were recognized to be important not to generate free amino acids, but for different biological purposes.

Studies on blood coagulation, fibrinolysis, and complement activation gave support to the concept of limited proteolysis, coined by Linderstrom-Lang, who set proteolysis as an important enzymatic player in the homeostatic processes, since inactive proteins are transformed into bioactive proteins and peptides [12].

Before the middle of last century, the lack of homogeneous proteolytic enzyme preparations presented a serious problem, since it did not allow the determination of enzyme specificity, i.e., to identify which proteinase or peptidase splitted a specific peptide bond. This problem was solved by J. H. Northrop, who achieved this in preparing crystallized enzymes [13]. His crystalline trypsin allowed Rocha e Silva to demonstrate that a bioactive peptide (bradykinin) was the product of a specific proteinase activity [14].

The specificity of proteolytic enzymes, the determination of the amino acid sequence of proteins and peptides, the synthesis of substrates and peptides, and the study of the mechanisms of catalysis required a fantastic effort from the pioneers of protein chemistry, such as Emil Fischer, Waldschmidt-Leitz, Max Bergmann, Joseph S. Fruton, Leonidas Zervas, J. H. Northrop, and many others. Particularly, it was after the development of the method which sped up the determination of amino acid sequence of proteins that the protein and peptide chemistry gained acceleration and dissemination throughout the world [12].

The bioactive protein include peptide hormones, neuropeptides, immunogenic peptides (epitopes), and others with no well-defined mechanisms. Only a few bioactive peptides have been discovered until the middle of last century, such as angiotensin, bradykinin, oxytocin, vasopressin, and insulin. This contrasts with the ever growing number of new peptides, which are constantly being added to the thousands of bioactive peptides already described.

·   ·   ·   ·

CHAPTER 2

Bioactive Peptides Produced by Extracellular Proteolysis

2.1 THE GASTROINTESTINAL SYSTEM (GIS)

Food proteins have a vital role in providing amino acids for the synthesis of our own proteins. However, partially or incompletely digested proteins may be assimilated, frequently as noxious peptides. Various symptoms of protein poisoning are experienced by different individuals, including burning in the mouth, lips and throat, the skin, nasal irritations, and other signs of intolerance to certain proteins, which can harm or even kill us.

Differently from other systems described in this book, the generation of bioactive peptides from food occurs by chance and not as an organized proteolytic processing of precursor proteins.

Proteolysis within the GIS is the most powerful machinery for protein and peptide degradation. In men, the GIS proteolytic system is able to carry out the digestion of 70–100 g of protein/ day [15]. Food proteins are degraded in the GI juice, into which a number of proteinases and peptidases of the salivary, the pancreatic, and the parietal glands are secreted, along with mucus and HCl. Besides the proteins from food, we find proteins derived from the large commensal microbial communities, whose participation in our food digestion and immunological protection is still poorly understood. Each of us harbor several hundreds of bacterial species in our GIS (trillions of microbes), which are integrated in our physiology by fully aggregating their own genomes (microbiomes) to ours, in a dynamic relationship that includes changes in bacterial composition according to dietary ingredients of the host, within and along their living conditions and evolution [16]. It was recently shown that the functional repertoire of microbiome genes is adjusted to the diets that feed distinct mammalian phylogenies or humans, living in a variety of habitats [17]. In addition, for the same gnotobiotic animal, the functional repertoire of microbiome genes varies according to the predominance of a specific component of the food (protein, carbohydrates, or lipids) [18].

According to many studies, the robust immune defense against non-digested food, or partially digested food, indicates how complex the immunity is in the gut [19]. The initial antigen exposure may generate a strong T cell-mediated suppression, and the so-called oral tolerance is developed. Otherwise, these proteins can cause a number of known diseases, such as IgE-mediated food allergy [20], as detailed by Vickery et al. [21].

The penetration of undesirable protein products or deleterious bioactive peptides into the internal tissue and fluids of the body is prevented by a single layer of epithelial cells of the intestinal mucosa, where these products are confronted with the largest antigenic challenge of any mucosal surface of the human body.

Protein products of commensal microorganisms are neutralized by signaling to the innate immune system, which, through pattern recognition receptors, responds to the threat by developing tolerance [22]. The commensal microbiota is important for several reasons, which include helping the GIS to digest poorly hydrolyzed proteins, such as mucus glycoproteins [23]. At the intestinal wall, brush border epithelial cells, rich in peptidases, contribute to complete the degradation of remaining peptides by hydrolyzing them into free amino acids (Figure 1).

In spite of the powerful machinery for protein degradation, a few peptides are able to resist the attack of proteinases and peptidases. Resistant peptides may result from cross-linking reactions, glycosylation, or high proline and glutamine content, which makes them resistant to hydrolysis. They usually are di- and tri-peptides, or medium-sized peptides that are resistant to enzymatic degradation. Some of these peptides can be absorbed by GIT mucosa cells and, subsequently, are able to produce biological effects [24, 25, 26].

How can food-derived peptides, preserved from this massive destruction, act as bioactive molecules?

In spite of the difficulties faced by peptides to overcome the barriers imposed by the intestinal mucosa, there is evidence in the literature suggesting that some peptides survive and are biologically active, even if they are absorbed only in minute amounts. Among the most-studied bioactive peptides originated from food, there are peptides displaying antihypertensive, opioid, anti-tumor, or autoimmune effects.

2.1.1   Antihypertensive Peptides

Some bioactive peptides, generated from food, have been shown to be beneficial. Milk, colostrum, and related products are among them [27]. A recent review article on this subject presents evidence suggesting therapeutic actions of milk peptides for several human pathologies. Milk-derived di- and tripeptides, formed by digestive enzymes, or by the proteolytic enzymes of lactobacilli, during the fermentation of milk can be taken up by intestinal epithelial cells in intact form. *In vivo* effects include opioid-like, metal-binding, antithrombotic, immunomodulatory, or angiotensin-converting enzyme (ACE) inhibition [28]. Some antihypertensive peptides from milk digestion, such as Val–Tyr, Ile–Tyr, Phe–Tyr, Ile–Trp, Val–Pro–Pro, and Ile–Pro–Pro, reduce arterial blood pressure in spontaneous hypertensive rats [29]. In fact, some of these anti-hypertensive peptides, administered by the oral route, were found to be present in the aorta [30], suggesting that the in-

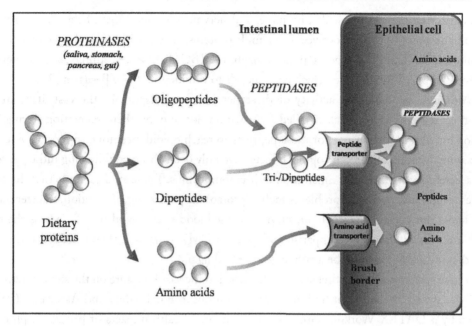

FIGURE 1: Gastrointestinal proteolysis. The gastrointestinal proteolytic system is responsible for processing of proteins from food ingestion. Such exogenous proteins are firstly degraded by proteolytic enzymes (proteinases) present in the saliva, stomach, pancreas, or gut. These enzymes cleave the proteins into oligopeptides. In the gastrointestinal system, there are also peptidases that cleave the peptides into smaller fragments or free amino acids. The di-/tri-peptides that resisted enzymatic degradation and the free amino acids are then transported to the intracellular environment through specific transporters present in the brush border epithelial cells. Once absorbed by gastrointestinal mucosa cells, the peptides can be converted into free amino acids by cytosolic peptidases or reach the circulatory system.

testinal peptide transporter allows their efficient delivery into the circulation. Accordingly, the association between milk consumption and healthy blood pressure was reported in the first National Health and Nutrition Examination Survey [29].

The anti-hypertensive activity of these di- and tri-peptides is associated with their inhibition of the angiotensin-converting enzyme (ACE) activity of endothelial cells [29, see next Chapter]. However, larger and proteolytically resistant proline-rich peptides, such as the teprotide, exhibiting stronger inhibition activity toward the ACE, did not present any anti-hypertensive effect when given by oral route, most likely because it is unable to cross the intestinal barrier [31].

2.1.2 Opioid Peptides

Opioid peptide sequences have been characterized in food proteins derived from animal and vegetable products. It has been suggested that they may contribute to an increased risk for certain diseases affecting the cardiovascular, immune, and nervous systems. Most of the studies have focused on the opioid peptides derived from milk proteins, in particular, the β-caseins known as β-casomorphins. The prototype is β-casomorphin-7 (BCM7), whose amino acid sequence is Tyr–Pro–Phe–Pro–Gly–Pro–Ile, which corresponds to residues 60–66 of β-casein [32].

Although the biological activity of these peptides is undisputable, the vast literature that covers their relevant aspects lack detailed information not only on their generation during dairy digestion but also on the chance for these peptides to reach opioid receptors. In most, if not in all, *in vivo* animal studies to date, the opioid effects have only been observed following intra-peritoneal or intra-cerebro-ventricular administration of β-casomorphins. These studies imply that the opioid peptides could easily overcome problems such as proteolytic processing, absorption, transfer mechanisms across the intestinal epithelium, survival in the blood stream, and transfer across the blood brain barrier. In addition, the comparison of bovine BCM7 to medicinal and endogenous opioids shows that it does not seem to be a very potent opioid ligand.

These and a number of other aspects, described in the vast literature on the subject, have been critically analyzed by the Scientific Report of the European Food Safety and Assistance (EFSA), prepared by a DATEX Working Group on the potential health impacts of β-casomorphins and related peptides [32, 33].

A better chance for β-casomorphins to exert *in vivo* biological activity seems to be the possible physiological role of these peptides in the intestinal lumen through direct stimulation of the opioid receptors of epithelial cells. It has been shown that the BCM-7 promotes a strong release of mucin in the jejunum of rats by activating the enteric nervous system and opioid receptors of epithelial goblet cells [34]. Obviously, this effect of BCM-7 could have dietary and health applications.

2.1.3   Cancer Preventing Peptides

Human clinical trials are being performed with Bowman Birk Inhibitor Concentrate from soy beans. Its cancer preventive activity relies on two major components, the 43-amino acid peptide lunasin and the 71-amino acid polypeptide Bowman Birk Inhibitor (BBI), which have been proposed to act together, lunasin being the bioactive cancer-preventing agent and BBI protecting it from digestion by trypsin and chemotrypsin of the GIS [35].

Lunasin's unusual structure, consisting of four domains, seems to be the clue to its mechanism of action in suppressing both chemical and oncogene driven transformation of mammalian cells. While the function of residues 1–22 is currently still unknown, residues 23–31 may target lunasin to the chromatin; the cell recognition sequence of residues 32–34 (Arg–Gly–Asp) might internalize lunasin into the cell nucleus, and the poly-aspartyl end of lunasin (residues 35–43) could be responsible for binding it to the core histones within the chromatin, thereby inhibiting their acetylation [36]. Importantly, it was established, that, after oral consumption, the peptide is absorbed, reaching target tissues and organs in an intact and bioactive form. It internalizes into cells and localizes to the nucleus [37, 38]. Differently from peptides derived from gluten (see below), lunasin is not generated by the enzymes of the GIS, but is ingested in its active form and protected from degradation by the soybean polypeptide BBI.

The cancer-preventing properties have been shown in a series of experiments in mice, in rats, and with human cell lines. Among others, results demonstrated that lunasin inhibits skin tumorigenesis in mice upon topical application [38], that it inhibits foci formation in a xenograft model of nude mice transplanted with human breast cancer cells, and delays the appearance of tumors [35]. Lunasin's properties of inhibiting cell proliferation of transformed cells and inducing their cell death make it a promising candidate for anti-neoplastic treatment.

2.1.4 Immunogenic Peptides

Being of upmost medical importance, an array of experimental evidences is available to demonstrate how peptides, derived from food proteolysis, produce noxious effects to the immune system. Obviously, the complete processing of food proteins is expected to abolish any chance for protein antigenicity. An additional mucosal defense is given by a hydrophobic layer of mucin oligosaccharides and the epithelial junction complexes, which are in charge of trapping antigens. Abnormalities in mucin secretion, or in the epithelial barriers, may be responsible for gastritis, bacterial overgrowth, and colitis [39, 40]. Consequently, in order to prevent the growing number of pathologies related to noxious or inappropriate human nutrition, it is necessary to better understand the relationship between diet, nutritional status, the immune system, and the GIT microbioma in humans, at different stages of life and in various habitats [16].

Incomplete hydrolysis of protein fragments may preserve or give rise to epitopes, thereby either triggering the immune tolerance mechanisms or causing food allergy [41]. After crossing the epithelial layer and interacting with immune defense cells (e.g., lymphocytes, antigen-presenting cells, dendritic cells and many other cells of the mucosa lymphoid tissue), food antigens stimulate immune tolerant T lymphocytes, the ultimate effectors of oral tolerance. However, a number of them are, or may be transformed into, noxious peptides by intracellular enzymatic processes. They are usually responsible for human diseases. A complex chain of reactions may occur as illustrated in Figure 2.

Increasing evidence suggests that the mucosal epithelium is likely to be more involved in tolerance than simply acting as a physical barrier. Epithelial cells are known to express MHC class II molecules on their basolateral membranes, and thus might act as nonprofessional APCs, which do not express conventional co-stimulatory molecules, favoring anergy [42]. In addition, factors derived from the gut epithelium are generally believed to condition the dendritic cells in the stroma, dampening immune responses and promoting gut homeostasis [43]. The cell membrane protects

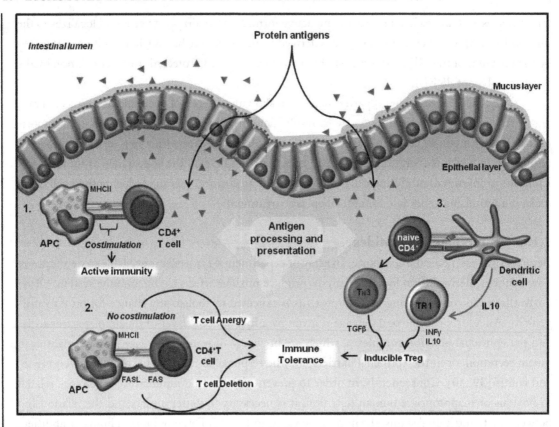

FIGURE 2: Peptides from protein antigens and their immunological implications. Dietary proteins are processed through the gastrointestinal tract and reach the intestinal lumen where their products can be absorbed by the epithelial cells. Peptides generated from food function as antigens (purple triangles), thus behaving as noxious bioactive peptides. The mucus layer and the epithelial junction complexes form the mucosal defense against such antigens, but exceptionally they can cross these barriers, be processed and presented by antigen presenting cells (APCs; dendritic cells, macrophages), and then interact with immune defense cells of the mucosa tissue. Such interactions result in different immune response, for instance, active immunity (1) or immune tolerance mechanisms (2 and 3).

living cells from the products of protein digestion, only allowing the controlled traffic of compounds across this barrier into the cell, even in the case of small molecules. However, short, water-soluble and partly hydrophobic, and/or polybasic peptides can cross the cytoplasmatic barrier as has been demonstrated for the cell-penetrating peptides (CPPs). CPPs consist mostly of 30 to 35 amino acids residues, and are able to penetrate the cell membrane at low micromolar concentrations *in vivo* and *in vitro* without requiring energy, any chiral receptor, or causing any significant membrane damage [44]. On the other hand, peptides with a high content of proline and glutamine can cross the GI barrier and cause pathological effects (see below).

Food allergy affects approximately 4% of the children and 2% of the adults in the US. Specific biochemical properties, common among different food groups (eggs, milk, soy beans, peanuts, tree nuts, and seafood) may confer resistance to enzymatic degradation. Collectively, they may confer allergenicity, when the fragments of these proteins reach the small intestine. Glycosylation, resistance to thermal denaturation, abundance in the food source, linearity of epitopes, solubility in water, and presence of prolines in the amino acid sequence are among the properties of the non-digested peptides.

One of the best examples of peptides, which overcome the obstacles of this fantastic GI barrier, are those derived from gluten degradation, responsible for Celiac disease, a highly prevalent immunogenetic enteropathy, caused by the ingestion of gluten. It is a polygenic disorder, and HLA-DQ2 is the single most important genetic factor [45]. Gluten-reactive T cells recognize peptides from gluten in the context of HLA-DQ2, but not in the context of any other HLA molecules expressed by patients [46]. This autoimmune-like dysfunction that affects 1:200 in many populations is the result of partial digestion of gluten composed of gliadins and glutenins, proteins exceptionally rich in proline and glutamine residues [47]. The most frequent epitopes, Gln–Pro–Phe–Pro–Gln–Pro–Glu–Leu–Pro–Tyr and Pro–Gln–Pro–Glu–Leu–Pro–Tyr–Pro–Gln–Pro, are derived from α-gliadin [25, 26, 48]. They are largely resistant to degradation by gastric and intestinal proteinases and peptidases [49]. The immunological reaction is mediated by HLA-DQ2 which has unique antigen-binding properties directed toward peptides derived from gluten, thus instigating an inflammatory reaction [50]. When genetically predisposed individuals, expressing the human leukocyte (HLA)-DQ2 dimers, bound to these immunodominant epitopes, are presented at the surface of APC, a robust response of the restricted CD4$^+$ T cells in the small intestinal mucosa of celiac patients is elicited, thereby inducing a sustained strong inflammatory response [51].

The unique ability of HLA-DQ2 to accommodate proline-rich peptides at the peptide binding groove seems to be very important to confer T cell stimulation by gluten peptides [50]. Interestingly, it has been shown that the binding motif for HLA-DQ2 includes the preference for an internal negatively charged amino acid within the peptide. The conversion of immunodominant epitopes, lacking an internal acidic amino acid, was resolved with the discovery that a ubiquitous tissue transglutaminase 2 could convert glutamine into glutamic acid, thereby introducing the negative charges required to bind to HLA-DQ2 [51].

In conclusion, the T cell-mediated inflammatory enteropathies caused by the ingestion of proteins capable of releasing immunoreactive epitopes could basically be envisaged as a shortcoming of the GI proteolytic system to inactivate any kind of noxious proline-rich oligopeptides. Many strategies have attempted to destroy the immunodominant epitopes in order to treat celiac disease.

These strategies range from inactivating transglutaminase 2 activity [52] to the introduction of bacterial prolyl peptidases to help digest proline-rich oligopeptides generated from food proteins [48, 53, 54]. Interestingly, no attempt has been described in the literature related to the use of the thiol-activated cytosolic prolyl-oligopeptidase as a possible agent to treat celiac disease. This ubiquitous serine peptidase hydrolyses internal prolyl residues of oligopeptides [55, 56, 57]. Consequently its activation could be useful to inactivate noxious intracellular proline-rich oligopeptides.

2.2   THE CARDIOVASCULAR SYSTEM (CVS)

Schoenheimer's findings had a large impact on biomedical sciences. One of the most fruitful discoveries was that proteins are subjected to degradation not only to supply amino acids for protein synthesis during starvation but also to provide bioactive fragments, essential for the physiological processes of living organisms. Theoretically, a bioactive peptide, buried within the structure of any protein, could be waiting to be released by the correct proteolytic system in order to exert its function. Among the huge number of examples, the first ones shedding light on the importance of this concept were those peptides generated in the cardiovascular system (CVS) by a chain of proteolytic events.

Bioactive peptides contribute to a large number of activities in the blood circulation. Blood provides nourishment and oxygen to tissues, eliminates catabolic products, promotes water and ion balance, and defends against injuries, among many others tasks. There is no need to describe the importance of the CVS in maintaining overall homeostasis.

The landmark studies on the relationship between bioactive peptides and the regulation of arterial blood pressure began in 1898 with Tigerstedt and Bergman, who suggested that a substance, present in kidney extract (renin), produced a vasopressor effect in rabbits [58]. Their pioneer contribution gained acceleration in the early 1930s, after the *in vivo* studies by Harry Goldblatt, who demonstrated that a reduction in blood circulation in the kidneys produced systemic arterial hypertension [59]. In the late 1930s, two research groups, working independently, discovered that the arterial hypertension, generated in the experimental model of Goldblatt, was due to the release of a substance (angiotensin) produced by renin. One of the laboratories was led by Dr. Bernardo Houssay at the University of Buenos Aires, Argentina, and the other by Dr. Irvine H. Page at Eli Lilly Research Laboratories in Indianapolis, USA. They discovered that the arterial hypertension of the CVS could be mediated by the renin–angiotensin system (RAS), when reduction of blood circulation in the kidneys occurred [60, 61].

Nine years later, Rocha e Silva and co-workers at the Biological Institute in São Paulo, Brazil, discovered bradykinin [14], a nanopeptide, that functions as the physiological antagonist of angiotensin II in the regulation of arterial blood pressure [62]. They arrived at this result while studying the hypotensive effect of the venom of the *Bothrops jararaca* snake on the blood circulation of dogs. The effect was supposed to be due to the action of putative proteolytic enzymes in the snake venom, since the hypotensive substance could be obtained *in vitro* by treating plasma with crystallized trypsin provided by Dr. J. H. Northrop. To our knowledge, this was the first time that a homogeneous proteinase preparation was used to demonstrate the importance of limited proteolysis in the release of a bioactive peptide [14]. These experiments were confirmed a few years by Elliot and colleagues [63].

Consequently, the precursors of the bioactive peptides and the proteinases, responsible for their physiological release had to be isolated and characterized. In the mid 1950s this project repre-sented a fantastic challenge for any laboratory. Before the advent of molecular biology, the complete characterization of the molecular participants of proteolytic reactions and their bioactive products required gigantic efforts from a number of skilled scientists and technicians [64]. For the amino acid sequence determination of proteins and peptides, the synthesis of peptides, inhibitors, and chromogenic and fluorogenic substrates, dozens of chemicals and technological devices still had to be developed; an endeavor that took 40–50 years. Notwithstanding, the identification and charac-terization of the endogenous precursors of the proteinases and peptidases responsible for the release of angiotensin II (Ang II) and bradykinin (Bk) progressed, allowing the description of coherent biosynthetic pathways that generate Ang II (RAS) [65, 66] and Bk (kinin–kallikrein system, KKS) [67] and explain their biochemical, physiological, or pathological interactions [62].

The limited proteolysis reactions that release the vaso-contracting peptide angiotensin II (Asp–Arg–Val–Tyr–Ile–His–Pro–Phe) and the vasodilator peptide bradykinin (Arg–Pro–Pro–Gly–Phe–Ser–Pro–Phe–Arg) into the circulatory system do not occur by chance. They involve an organized and complex process of enzyme–substrate interactions, occurring at the epithelial cellular plasma membrane or at specific locations within the vast compartment of blood vessels. If limited proteolysis would occur at a non-specified region of the blood vessels, the released bioactive pep-tides would not survive long enough to be able to interact with their receptors and elicit contraction or dilation of the arterial vessels.

Moreover, the particular step of limited proteolysis that leads to the release of Ang II or Bk occurs during a complex chain of proteolytic events involving plasma proteins, blood cells, the kidneys, and endothelial cells. The stage is the circulatory system, and the reactions represent a re-sponse to various needs (blood clotting, fibrinolysis, angiogenesis, regulations of fluids, ions, arterial blood pressure, response to injuries), all together contributing to the homeostasis of the cardiovas-cular, the renal, and the defense systems [68]. This panoramic view shows clearly that these chains of proteolytic events are not at all similar to the digestive proteolysis occurring in the GIS (see previous section). As a whole, the limited proteolysis processes, occurring in the blood, are highly selective and include activation of two distinct enzymes: (i) specific proteinases located nearby their substrates that perform the initial cleavage of the protein and (ii) peptidases that are required to conclude the maturation of a given bioactive peptide in order to give rise to proper ligands whose structures are able to be recognized by the nearby selective receptors. Together, these two steps produce the bioactive peptides that maintain blood circulation homeostasis.

Concerning the release of bioactive peptides *in vivo*, a number of properties of the substrates and the proteolytic enzymes are critical for the successful extracellular limited proteolysis. Important factors include the specificities of proteinases and peptidases, the location for processing or trimming, the accessibility of the susceptible peptide bonds, and the modulation of the enzyme activities by natural inhibitors and activators. Because of this complexity, it is difficult to be sure about the physiological or pathological significance of the results obtained *in vitro*, *ex vivo*, or in experimental animal models (including knock-out animals) for the release of a specific peptide.

In this chapter, we will focus on the limited proteolysis that gives rise to bioactive peptides participating in physiological and/or pathological events of the CVS. It is not our purpose to provide details on the physiopathology of the RAS and the KKS, since they are too complex, beyond our focus, and still subject of intensive studies. Several review articles will be suggested that cover the particular subject.

2.2.1   The RAS

Angiotensinogen (the substrate of renin) is a 60-kDa glycoprotein found in the α_2 globulin fraction of plasma proteins, synthesized and released in the liver. It is cleaved in the vascular bed to generate the biologically inactive Ang I (Asp–Arg–Val–Tyr–Ile––His–Pro–Phe–His–Leu), a decapeptide cleaved off from the N-terminal portion of angiotensinogen by renin, an aspartyl proteinase synthesized as prorenin. The physiological maturation of prorenin into active renin takes place at the juxtaglomerular cells of the kidneys, where it is secreted into the blood [69]. Renin and prorenin secretion is stimulated by the decrease in blood flow and blood pressure, and the loss of Na^+ and water [62, 67].

The limited proteolysis reaction leading to the release of Ang I occurs at the prorenin-receptor, located at the cytoplasmic membrane of the vascular endothelium or the smooth muscle cells, or within the extracellular matrix, where prorenin is converted into renin. Alternatively, circulating renin may bind to the prorenin-receptor, promoting the selective excision of Ang I from angiotensinogen in a rate-limiting step. The membrane-bound renin is five times more efficient than renin in solution [65, 66]. The existence of a receptor is not unique to this proteinase; other proteinase receptors have been described [70].

The decapeptide Ang I needs to be converted into its bioactive form, angiotensin II (Ang II; Asp–Arg–Val–Tyr–Ile–His–Pro–Phe), a reaction that should occur in the vicinity of the Ang I release. The processing of Ang I into Ang II requires the removal of 2 amino acids from the C-terminus, a step that is mediated by angiotensin converting enzyme (ACE).

After the conversion, Ang II specifically interacts with its receptor (AT1R), which is a G protein-coupled receptor (GPCR). The distribution of ATR1 is ubiquitous and abundant in adult tissues. Once activated, ATR1 elicits a wide spectrum of physiological actions, such as increase of blood pressure, thirst, and sodium demand. In addition, ATR1 activation can cause a large variety of pathological actions in the cardiovascular and renal tissues [71]. Ang II can also interact with an alternative GPCR (AT2R) found at high concentrations in the fetus. Although present at low concentrations in adults, this receptor has been described to be up-regulated in pathological conditions [72].

In 2002, another ACE-like enzyme was found in the heart, a carboxypeptidase [73] that converts Ang II into Ang $_{(1-7)}$, a hypotensive peptide of the RAS, resulting from the removal of the C-terminal phenylalanine of Ang II. The enzyme is known as angiotensin-converting enzyme 2 (ACE2). Ang $_{(1-7)}$ activates Mas, its specific GPCR, which counteracts the hypertensive effects of Ang II in the long-term regulation of body fluids and arterial pressure [74]. ACE2 is mainly expressed in the heart, the kidney and the testis.

2.2.2 The KKS

More complex reactions involve the physiological release of Bk (Arg–Pro–Pro–Gly–Phe–Ser–Pro–Phe–Arg) from kininogen by the kalikrein-kinin system (KKS). The classical mechanism occurs in the blood circulation during several physiopathological processes, but might as well occur in hemostasis [68]. Hemostasis is a crucial process that, ultimately, preserves the volume of circulating blood in the organism. A proteolytic cascade, triggered by tissue damage, involves the participation of several plasma proteins, blood, and endothelium cells. A chain of proteolytic activities results in blood clotting and in the release of Bk into the biophase of the endothelium cells. Limited proteolysis releases Bk from high-molecular-weight kininogen (HK). The proteinase responsible for this reaction is plasma kallikrein, a serine proteinase secreted by the liver plasma prekallikrein (PK), a zymogen that also needs to be activated. HK is an inactive protein of 626 amino acids and a MW between 88–120 kDa, depending on glycosylation. The secreted HK quickly adsorbs to the cytoplasmic cell membranes of endothelial cells, an important step to confine the proteolytic reaction of the release of Bk near its receptor. After binding, the HK binds circulating PK, or alternatively, it might already be conjugated to PK; subsequently, the plasma PK is converted into kallikrein, the active proteinase. This activation requires the contact of PK with a negatively charged surface (platelet thrombus, for instance), whose mechanism remains elusive [75]. Once activated, plasma kallikrein reacts with HK and releases Bk. The activation of KKS is important for many other activities, such as thrombosis, urokinase activation, anti-angiogenesis, and anti-proliferation of cell growth [67, 76]. Bk and Lys-Bk can also be released by another enzyme, tissue kallikrein 1, which is highly expressed in kidney. This enzyme is a member of the tissue kallikrein family.

Once released, Bk is rapidly metabolized by endothelium peptidases including angiotensin-converting enzyme (ACE), neutral endopeptidase (NEP), carboxypeptidase N (CPN), and aminopeptidase P [77]. Bk has a half-life of 15 seconds, and circulating levels are usually relatively low (0.2–7.1 pM) [78].

The biological effect of Bk results from its interaction with a specific GPCR (B2R), which is constitutively expressed and subject to rapid desensitization. Binding of Bk promotes potent hypotension [79]. Bk also participates in inflammatory processes by activating endothelial cells to promote vasodilation and increased vascular permeability, producing the classical symptoms of inflammation such as redness, heat, swelling, pain, and attraction of immune cells. Specifically, Bk contributes to tissue hyper-responsiveness and local inflammation in allergic rhinitis, asthma, coughing, and anaphylaxis [80].

In addition to the ubiquitous B2R, another Bk-receptor exists, the inducible B1R. However, full-length Bk does not activate this receptor directly and requires conversion into des-Arg-9 Bk by carboxypeptidase N or a related enzyme. BR2 is important in inflammatory processes.

Limited proteolysis is directly responsible to responding to injuries leading to inflammatory processes. It participates in several steps when a plethora of cell-derived mediators (e.g., chemokines, cytokines, antimicrobial peptides, or reactive oxygen, and nitrogen species) are released, and KKS activation occurs in the vascular compartment. For instance, one important player of the inflammatory process is interleukine-1β (IL-1β), a 17.5-kDa cytokinin released by caspase I after hydrolysis of Asp^{116}–Ala^{117} of pro-interleukine-1β [81, 82]. This cytokinin is mainly produced by myeloid cells such as macrophages, monocytes, and dendritic cells. It leads to the recruitment of a variety of immune cells which act in concert with local tissue cells to remove the infectious agent, clear damaged tissue components, and initiate tissue repair processes [82]. It has been shown that IL-1β leads to up-regulation of B1R [83, 84]. Therefore, the hydrolysis of Arg^9 by the enzyme carboxypeptidase M at the cytoplasmic membrane of the endothelium converts Bk into des-Arg-9 Bk which activates B1R receptors, eliciting inflammatory response including pain [85, 86, 67]. Figure 3 summarizes the proteolytic activities in the blood, involving the RAS and the KKS (Figure 3).

In addition, bioactive peptides from RAS and KKS can be subjected to further hydrolysis to give rise to several smaller bioactive peptides whose participation in physiological or pathological events have not yet been conclusively demonstrated [67, 76].

2.2.3   Interactions between RAS and KKS through Bioactive Peptides

The best-defined molecular interaction between Ang I and BK was revealed by Yang and co-workers in 1970. They demonstrated that the same peptidase, now known to be ACE, was responsible for the conversion of Ang I into Ang II, and for the inactivation of BK [87]. This fundamental finding

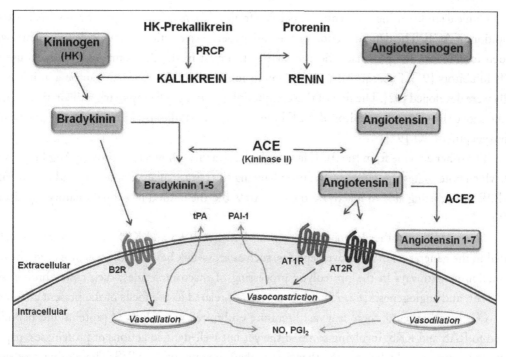

FIGURE 3: Crosstalk between reninn–angiotensin system (RAS) and kallikrein–kinin system (KKS) Prolylcarboxypeptidase (PRCP) can convert HK-prekallikrein to kallikrein. Kallikrein digests kininogen (HK—high molecular weight kininogen) to release Bk and also converts prorenin to renin. Renin has the ability to convert angiotensinogen to Ang I. ACE (kininase II) converts inactive Ang I to the vaso-constrictor Ang II and degrades Bk into Bk $(1-7)$ (not shown) or Bk $(1-5)$. Both angiotensin-converting enzyme 2 (ACE2) and PRCP can degrade Ang II to the vasodilating peptide Ang $(1-7)$ that stimulates nitric oxide (NO) and prostaglandin I2 (PGI_2) formation, which potentiates the effects of Bk. Ang II stimulates plasminogen activator inhibitor 1 (PAI1), released from endothelial cells, while the fragment of Bk stimulates the release of tissue plasminogen activator (tPA), inducing a thrombin inhibitory activ-ity. Therefore, the PRCP-degradation of the vasoconstrictor Ang II leads to the increased formation of the vasodilators Bk and Ang $(1-7)$. B2R, a constitutively expressed Bk receptor; AT1R and AT2R, constitutively and induced Ang II receptors, respectively.

led John Vane (Nobel Prize Laureate 1982) to suggest the development of an inhibitor of ACE for the treatment of human hypertension [88]. The search for inhibitors of ACE led to the discovery of naturally occurring proline-rich oligopeptides (PROs) in the venom of *Bothrops jararaca* [see 89]. One of these peptides has a strong preference for the C-site of the ACE, where most of the Bk inactivation occurs [90].

Clinical studies using one synthetic peptide from *Bothrops jararaca*, teprotide, revealed potent inhibition of ACE. The demonstration of the anti-hypertensive action of teprotide in humans was needed to convince the pharmaceutical industries to invest in the development of clinical use of ACE inhibitors [91]. Subsequently, highly potent inhibitors of ACE which can be administered orally were developed [92]. The first of these, captopril, was designed employing a theoretical model for the active site-directed inhibitor of ACE based on similar studies performed with the pancreatic carboxypeptidase A1 [92].

The other enzyme from the ACE family, ACE2, was also shown to hydrolyze Ang I into Ang 1-9, which potentiates the action of Bk, by enhancing the release of aracdonic acid and re-sensitize the B2R of Bk, being able to inactivate the des-Arg9-Bk, the natural pro-inflammatory agonist of B1R [93].

There are many other layers of interactions between RAS and KKS which are not directly related to the generation of bioactive peptides, such as cross-talk between the receptors [94], or the interaction of pathways in the proteolytic processing of macromolecules, such as in thrombosis, fibrilolysis, and angiogenesis [67, 75, 76]. They are not related to the focus of the present chapter.

On the other hand, there is a vast literature on limited proteolysis of proteins and peptides related to RAS and KKS, involving some of the yet not well-defined actions of proteinases, peptidases, and receptors, which occur in other tissues than the ones of the CVS. These other proteases generate bioactive peptides in "local" as opposed to the systemic RAS and KKS. It has been shown that not only the kidneys can express molecules of RAS but also many tissues express RAS molecules including the brain, the adrenal gland, the pituitary gland, reproductive tissues, the gastrointestinal tract, hematopoietic tissue, the heart, and blood vessels. This broad expression suggests the presence of functionally active RAS in these tissues [95]. The resulting molecules display distinct activities, and their physiopathological roles are still objects of intense investigation. Accordingly, these "new" bioactive peptides may have similar properties as the classical progenitor molecules, but their receptors have not yet been identified and may possess synergistic, antagonistic, or completely unrelated biological effects.

Finally, the overall actions of bioactive peptides of RAS and KKS in a given organism seem to be determined by specific proteolytic pathways responsible for the release of these peptides when the locality and the time do require. Although each bioactive peptide acts on its specific receptor, the multi-functionality of Ang-I and Bk is greatly expanded by the action of peptidases on the physiological generation of Ang I- and Bk-related peptides through distinct receptors of different tissues or cells, eliciting distinct actions. In other words, we may conclude that the biological significance of a single precursor protein (or gene) for RAS and KKS is greatly broadened when the protein is subjected to a variety of proteinases and peptidases of the limited proteolysis systems, according to the physiological or pathological stimuli.

ACE and prorenin isoforms have been found in invertebrate to mammalians and homologies among proteins and peptides of RAS and KKS give support to an evolutionary view of the conservation of Ang II- and BK-related peptides. The full sequence of Bk was found in insects, and the peptide is probably released from its precursor by similar processing enzymes as in mammals. However, it appears that the physiological effect of insect Bk is distinct from the role this peptide plays in mammals [96, 97, 98].

.   .   .   .

CHAPTER 3

Bioactive Peptides Generated by Intracellular Proteolysis

As described above, Schoenheimer [99], Linderstøm-Lang [9], and de Duve [8] revolutionized the way proteins should be conceived, showing an area of dynamic equilibrium between synthesis and degradation of proteins. Intra- and extracellular proteolysis regulate nutrition, availability of water, salts, O_2, and nutrients to the tissues, as well as receptor-mediated processes, signal transduction, the cell cycle, reproduction, sex behavior, body development, embryogenesis, differentiation, immunological surveillance, inflammation, transcription regulation, quality control of proteins, and modulation of diverse metabolic processes, all essential to homeostasis.

In contrast to the large compartments of the GIS and CVS, where the extracellular proteolytic systems take place, the intracellular proteolysis occurs inside the microscopic volume of the cell. The cytosol contains tightly regulated proteinases, multi-proteolytic complexes (proteasomes, for instance), and peptidases either inside or outside of complex organelles (i.e., lysosomes, endoplasmatic reticulum, mitochondria, secretory granules).

However, the physiological role of the proteolytic systems is not restricted to total protein fragmentation in an arbitrary fashion. Specific fragments, once released from a protein, might display biological activity. This observation led to the concept of limited proteolysis, introduced by Linderstøm-Lang [9]. It represents another major contribution for the understanding of the physiology of living organisms, since it explains how the proteolytic process is able to disclose the biological activities buried within the structure of newly synthesized proteins, by means of an organized process of protein processing, generating bioactive fragments.

Concerning intracellular proteolysis, this book aims at presenting an integrated view of the actions of proteinases and peptidases, which raise considerable interest in the biomedical sciences. These enzymes have been demonstrated to participate in almost all regulatory processes subjected to physiological control, immune defense and surveillance, and neuro-endocrine regulation. On the other hand, we know now that dysfunction of protein structures affects cellular physiology, leading to serious pathologies [3]. The loss of functionality caused by unfolding and/or abnormal aggregation of protein molecules may condemn these proteins to degradation in order to avoid cell death [12].

3.1 THE NON-SECRETORY SYSTEMS

3.1.1 The Lysosome/Endosome System (L/ES)

In his first description of the lysosome, Christian de Duve in the mid-1950s showed that this cellular vacuolar structure contains cathepsins, proteolytic enzymes which degrade proteins not only into bioactive fragments but also down to their constituent amino acids [8]. These reactions occur in the acidic lumen of the organelle, where endogenous and endocytosed proteins penetrate and are degraded. A complex membrane confines these enzymes within the lysosome/endosomes, thus providing an essential internal acidic medium for the hydrolytic activity of cathepsins. This confinement is required to prevent cell proteins from massive degradation. The vesicles can occupy several locations inside the cell to regulate diverse responses to starvation and subsequent nutrient replenishment [5] (Figure 4).

The intracellular proteolysis by lysosomes raises several questions, such as which mechanism(s) regulates the translocation of proteins into the lysosomal lumen in order to be degraded? In fact, the extremely different half-lives of cellular proteins [100], varying from minutes to days, the distinct responses of different populations of proteins to lysosomal proteinase inhibitors, and the proteolysis that occurs in cell fractions that do not contain lysosomes [101] are some of the many arguments for a hypothesis including lysosomes as one among other processes of intracellular protein degradation. Although the proteolytic process itself is exergonic, a number of proteolytic processes require energy, prior to peptide bond hydrolysis (such as unfolding of proteins and transport of proteins into the lysosome) [102, 103].

By challenging the lysosome hypothesis, Aaron Ciechanover, Avram Hershko, Irwin Rose, and co-workers conducted a research which ultimately led to the discovery of the importance of ubiquitin, and how this small protein present in all cells is able to direct intracellular proteins to degradation by the proteasome, thereby originating the ubiquitin–proteasome system (UPS) [4, 104]. This discovery gave the appropriate answer to most of the open questions listed above. More importantly, it introduced intracellular UPS among the basic metabolic reactions of cells involved in a large number of essential physiological processes found in all eukaryotic cells. The details of the historical and experimental steps that led to the discovery of intracellular peptide generation by the UPS can be found in recent reviews [104, 105]. Ciechanover's work was acknowledged by the award of the 2004 Nobel Prize in Chemistry.

3.1.2 The Ubiquitin/Proteasome System (U/PS)

This protein complex contains a 20S barrel shaped sub-unit (about 700 kDa) consisting of four rings, arranged in a stack defining an internal multicatalytic chamber, where proteolysis occurs. The 20S proteasome includes 28 subunits which are products of two homologous genes (α and β). The β-subunits, containing threonine at the N-terminal, are essential for catalysis [106].

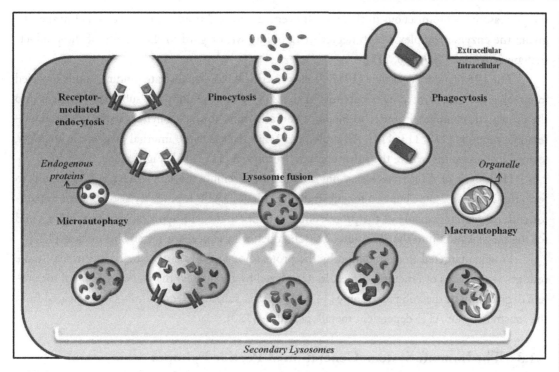

FIGURE 4: Lysosome proteolysis. Different types of endocytosis processes lead endosomes containing proteins, particles or organelles to lysosomal fusion. Receptor-mediated endocytosis; pinocytosis (engulfment of extracellular fluid); phagocytosis (engulfment of extracellular particles); microautophagy (engulfment of intracellular proteins); and macroautophagy (engulfment of oganelles) are represented in the figure where the white arrows indicate that each endosome/vacuole is targeted to the lysosome fusion process, and its contents are submitted to proteolytic processes in secondary lysosomes. The lysosomal/vacuolar system is a complex vesicle network, responsible for intracellular traffic and/or degradation of exogenous/endogenous compounds. Other components of this network are not shown in the figure.

In eukaryotes, the 20S proteasomes consist of seven β-subunits that are organized in six proteinase centers, exhibiting distinct substrate specificities [107, 108]. The β1 subunit hydrolyzes peptide bonds to the C-terminal side of negatively charged amino acid residues (Glu and Asp) and is often referred to as caspase-like, although this is incorrect because caspase is specific for Asp residues and does not cleave after Glu residues. The β2 has trypsin-like activity, hydrolyzing peptide bonds to the C-terminal side of positively charged amino acid residues (Lys, Arg). The β5 subunit has a chymotrypsin-like activity, preferentially hydrolyzing peptide bonds to the C-terminal side of hydrophobic amino acid residues (Leu, Phe, Tyr, and others). However, the cleavage site usage by the proteasome is promiscuous in that almost every amino acid residue can serve as a cleavage site, giving the enzyme complex a high degree of flexibility with regard to the quality of the products that are generated [109, 110, 111].

The 19S regulatory particle (19SRP) plays a critical role in substrate binding and is involved in opening the gate formed by α-subunits of the 20S, where the unfolded substrate advances into the proteolytic chamber, in order to be processed according to the catalytic selectivity of the proteasome subunits [112, 113, 114]. All proteinase active sites face the internal proteolytic chamber, formed by β-subunits, where the substrates are hydrolyzed [115, 116] (Figure 5).

Hydrolysis of ATP is necessary not only for the assembly the 19S with the 20S subunits to form the 26S proteasome but also for substrate degradation. ATP is probably also required for conformational rearrangements of the proteasomal subunits that occur upon substrate unfolding, and for the translocation into the 20S proteolytic chamber, whose mechanism remains unclear [117].

Proteins targeted to degradation by the proteasome are covalently labeled at internal lysine residues with chains of ubiquitin molecules (catalyzed by E3 ligases) that serve as a signal for protein degradation by the 26S proteasome [118, 119]. After substrate degradation, Ub is released from the substrate by an ATP-dependent metallopeptidase [120].

3.1.3 The Immune System: Generation of Peptide Epitopes

The work published in 1985 by Babbitt et al. [121] describes that the fundamental role that peptides derived from antigens (non-self proteins) have in immune defense. "These scientists found that a major component of the T-cell response was directed toward determinants found in the unfolded protein, and that the determinant presented to T cells was a sequence of only 10 amino acids" [121]. Subsequently, the new paradigm received robust experimental support [see 122]. Accordingly, immunoactive peptides became the focus of one of the most relevant areas of modern immunology.

The effectiveness of immune surveillance of intracellular pathogens (viruses, bacteria) and cancer/neoplasia depends upon a functioning antigen presentation pathway. The antigen-presenting cells (APCs) allow infected cells to reveal the presence of an intracellular pathogen, because once inside the cell, antigens are degraded by the L/ES or by the UPS, producing antigenic peptides that interact with the Major Histocompatibility Complex (MHC) which chaperones them to the cell surface [123]. These antigens are thereby exposed by the APCs. Peptides derived from the L/ES or from the UPS or from non-conventional sources bind to the MHC class II or MHC class I in the endosome of the cell [124].

Limited proteolysis promoted by cathepsins of the L/ES and by proteasomes of the UPS produces the peptides responsible for tolerance against self-destruction and for activation of the immunological system, turning the organism capable of clearance of pathogens and infected cells. For instance, to face a virus infection, antigenic peptides along with dozens of immunologically inactive

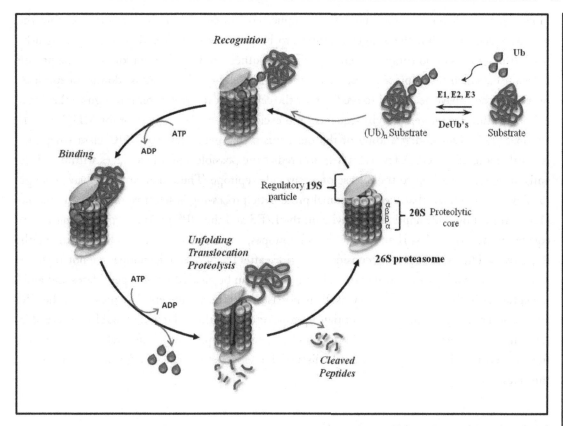

FIGURE 5: Ubiquitin-dependent degradation of proteins by proteasomes. The proteasome is a protein complex responsible for the degradation of ubiquitinated proteins through an ATP-dependent mechanism. It is composed of a proteolytic core and regulatory particles. The most common proteasome, the 26S proteasome, consists of a core called 20S particle and two 19S regulatory particle that recognize the polyubiquitinated proteins. A substrate (protein) is targeted to the proteasome degradation after conjugation of a poly-ubiquitin (Ub) chain (polyUb) to the protein substrate by enzymes (E1, E2, E3). This ubiquitin extension process can be reversed by a deubiquitinating enzyme (DeUb). Once labeled with polyUb, the protein ((Ub)n substrate) is recognized and binds to the proteasome. The protein is unfolded, translocated and cleaved by proteolytic activity of the 20S core originating oligopeptides.

fragments, are released when virus antigens are processed by the L/ES or by the UPS. The question still waiting for a proper answer is whether the proteolytic fragmentation of the antigen is carried out in an arbitrary or in an organized fashion.

Most existing data suggest that the immunization process depends upon the selection of immunogenic peptides, thus reflecting the development of a tightly regulated proteolytic process.

However, we still poorly understand the complicated proteolytic mechanisms that enable the epitope sequence within the antigenic protein to be precisely excised and selected, distinguishing it from dozens of non-immunogenic peptides. In other words, we do not know yet the precise mechanism, which allows proteinases, such as the L/ES and the UPS, to release dominant epitopes. Several methods have been used to predict how the cellular proteolytic systems recognize the linear epitope within the antigen protein sequence, and excise it. Results are still poor for MHC-class II epitopes [125]. Our understanding of the mechanism that generates the MHC class I peptides is a little more advanced. A few rules help to predict the possible sites of cleavages within a given antigen protein that lead to the release of a particular epitope. These data are conclusive enough to allow the statement that the proteasomal proteolytic processing is selective rather than random [126]. At least one rule appears to be evident: the L/ES and the UPS provide peptides lengths adequate for the MHC class II and MHC class I epitopes, i.e., about 15-24 aa and 8-12 aa in length, respectively. This is an important consideration since after being released, epitopes are not anymore susceptible to hydrolysis by proteinases, although they can be cleaved by oligopeptidases and aminopeptidases in the cytosol. These cytosolic peptidases, and the aminopeptidase present in the ER, either convert a peptide into an appropriate epitope or destroy them. How do the APCs succeed in selecting the immunogenic peptide in the middle of such an apparently unfavorable environment, filled with competitors and peptidases? As discussed in Chapter 4, further work is needed to answer this question.

3.1.3.1 Peptides and the MHC Class II

One of the best-known bioactive peptides produced by the lysosomal/endosomal system is the MHC class II epitope. These molecules are found only on professional APCs, which include macrophages, dendritic cells, and B-cells. They are responsible for the humoral immune response.

MHC class II epitopes are generally peptides of 15 to 24 amino acid residues in length found in the cytosol, as result of the degradation performed by cathepsins of the L/ES. They bind to MHC class II molecules that migrate to the surface of the APCs. Subsequently, they activate CD4$^+$ T cells leading to the expansion of CD4$^+$ effector-cells.

A bioactive peptide participates at the starting point of the MHC class II immune defense. During the biosynthesis of MHC class II molecules, an invariant protein chain joins the MHC class II molecules in the lysosome/endosome compartments. This invariant glycoprotein is subsequently hydrolyzed by endosomal cathepsins, releasing a peptide named class II-associated invariant chain, CLIP, (LPKPPKPVSKMRMATPLLMQALPM) [127], which binds to the antigen-binding groove of the class II receptor. This peptide is removed in order to allow the penetration of the MHC class II epitope into the groove of the MHC class II molecule, which subsequently migrates to the cell surface of APCs carrying the peptide in order to be recognized by the receptor T-cell helper (Th), thus eliciting the appropriated immune response by the CD4$^+$ effector-cells [see 128, 129].

3.1.3.2 Peptides and MHC Class I

In higher eukaryotes, the vast majority of intracellular self- and non-self-proteins (80–90%) is fragmented by the 26S proteasomes, generating countless oligopeptides, few of which might turn into an immunogenic peptide [130]. Exhibiting such features, how could the proteasome be an appropriate proteolytic system to generate a small set of immunogenic peptides to provide effective immune surveillance?

Cancer cells, bearing mutated proteins or infected cells, containing viral or bacterial proteins produce a new repertoire of oligopeptides. The presence of these foreign MHC class I peptides allows the CD8 T cells to specifically recognize, and, eventually, eliminate the infected or cancer cells. Thus, effective immune surveillance depends upon the cells' ability to generate a diverse MHC class I peptide repertoire via the antigen processing pathway. The diversity of peptides presented by MHC I is very broad, depending on the animal species. In humans, the number of MHC class I alleles is higher than 2,000, and every one of those may present a set of peptides of 8–10 amino acid in length that share a consensus motif defined by two or three conserved amino acids (hot spots). Thus, theoretically, billions of sequences can be presented [131]. The sequences of over 24,000 MHC class I epitopes have been deposited in a database [132].

The proteasome seems to be perfectly adequate for MHC class I antigen presentation; it selectively produces peptides with an average of 8–10 amino acid residues of length which are endowed with the appropriate amino acids required for binding to the MHC class I groove [131].

All peptides generated by the proteasome are potential bioactive molecules. Interestingly, recent work demonstrated that the proteasome of immune cells is distinct from the proteasome of the non-immune cells. The most profound influence on the notion that immunoproteasomes are involved in antigen processing resulted from the finding that, upon induction by immune stimuli, specific catalytically active β-subunits were incorporated into nascent proteasomes to form 20S proteasome complexes with an alternative subunit composition [133]. These subunits preferentially cleave at the carboxy-terminus after hydrophobic or basic residues (usually the preferential C-terminal anchor residue of MHC class I epitopes). Two of these subunits are encoded within the MHC class II region, which led to the terms immunosubunits (i-subunits) and immunoproteasome (i-proteasome) [134], implying that these enzyme complexes are responsible for the generation of a special set of peptides for antigen presentation.

Thus, the immunoproteasomes emerged as the main MHC class I generating epitopes because (i) they preferentially cleave at the carboxy-terminus after hydrophobic or basic residues (usually the preferential C-terminal anchor residue of MHC class I epitopes); (ii) proteasomes generate peptide sizes suited for binding to MHC class I molecules (8–10 amino acid residues); (iii) specific proteasome inhibitors abolished MHC class I antigen presentation almost completely [135].

Following the production of the peptides by the proteasome, a fraction of the peptides are transported into the lumen of the endoplasmic reticulum (ER) and bind to MHC class I molecules, which are then transported to the cell surface once peptide is bound. Transport across the membrane of the ER is mediated by TAP (transporter associated with antigen presentation), two related transporters located in the ER membrane. Recognition of the immunodominant epitope depends on its binding in the groove of MHC class I molecules [136, 137]. Some of the peptides generated by the 26S immunoproteasome are of the appropriate size and sequence to bind to TAP, as well as for docking to the peptide binding groove of the MHC class I molecule [135, 138, 139]. In some cases, peptides too long for MHC-binding are transported by TAP and trimmed by an ER aminopeptidase until the correct size for binding to MHC (see Chapter IV).

Cell surface-expressed MHC class I molecules present antigenic peptides which are specifically recognized by the T-cell receptor (TCRs) or by a coreceptor (CD8) of cytotoxic T lympho-

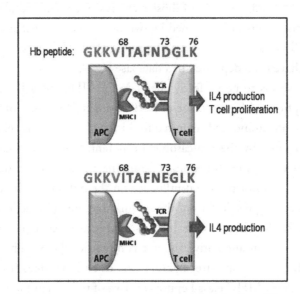

FIGURE 6: Influence of epitope structure on its immunological effect. The 64–76 hemoglobin (Hb) β-chain fragment is represented as a linear sequence, showing in green, the amino acid residues that interact with the TCR or the MHC. The change of one amino acid at position 73 (aspartic acid → glutamic acid, in red) modifies the peptide's function. This substitution does not affect the production of interleukin 4 (IL-4), but abolishes the T cell proliferation. APC: antigen presenting cell. Figure from reference [142].

cytes (CTL). These specialized T cells can detect other cells that endogenously express foreign or aberrant (i.e., mutated) protein molecules, and subsequently, remove these unwanted cells from the body [see 140].

Finally, the importance of the precise processing and/or modifications of the antigen should be strongly emphasized, since any changes are likely to affect the structural features of epitopes and, consequently, disturb their biological activity. Each peptide adopts a different conformation, while bound to the MHC molecule. Even conservative changes in the peptide molecule may cause abnormal immunological response [141]. In this case, the authors used the peptide Hb [64–76] from the β-chain of hemoglobin bound to MHC (pMHC). This complex was recognized as a T cell epitope by mice bearing the *Hbbs* variant. The interaction of MHC with the CD4-helper receptor elicited two kinds of response: production of IL-4 and in increase of Th cell proliferation (Figure 6).

The conservative substitution of aspartic acid [73] by glutamic acid did not affect the production of IL-4, but abolished the proliferative effect. Thus, this seminal experiment demonstrated the crucial importance of the epitope structure for its immunological effect. In addition, this study revealed the dissociation between two effects of a particular epitope by a simple modification of the epitope structure such as the removal of one methylene group of the peptide. This example opens up a great opportunity for research and applied science [see 142].

3.2 THE SECRETORY SYSTEM: THE NEUROENDOCRINE SYSTEM (NES)

One of the most fascinating episodes of medicine is related to the history of neuroendocrinology. Among a large number of books and articles on this subject, one publication is particularly interesting and was presented by Modlin and collaborators [143]. They describe the turmoil of concepts and ideas constantly modified by a large number of eminent scientists who contributed to a better understanding of the interplay occurring among cells, tissues, and biochemical events essential for systemic homeostasis. As the authors point out, "the contemporary neuroendocrinology should be conceived as a dynamic web of connections between the neuroendocrine and immune systems via the secretion of ubiquitous messengers that include peptides, amines and steroids, whose complexity leave the impression that we are constantly opening a Pandora's Box."

Our focus here concerns exclusively the involvement of bioactive peptides in neuroendocrinal processes. The regulation of homeostasis depends upon the actions of these molecules, which broadly act as endocrine, paracrine, neurocrine, and autocrine mediators. The peptides are products of limited proteolysis occurring within the cells, sharing a common phenotypic program characterized by the presence of neuropeptide precursors and processing enzymes, and dense core secretory granules that reach into the 'neuroendocrine-immunology' area [144, 145].

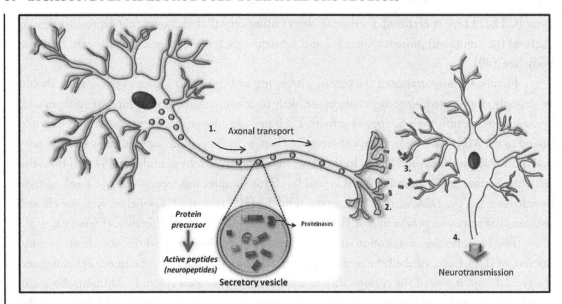

FIGURE 7: Formation of a neuropeptide. Neuropeptides (active peptides) can be generated from proproteins (protein precursor) by the action of processing proteolytic enzymes. Proteinases and peptidases cleave specific peptide bonds, giving rise to a single neuropeptide, multiple distinct neuropeptides, or multiple copies of a single neuropeptide. The neuropeptides are stored in secretory vesicles, transported through the axon (1) and released into extracellular space either at the synapse junctions or at non-synaptic sites (2). Once in the interstitial space, the peptides (green rectangles) can interact with their receptors (in red) (3), inducing a specific cellular response, for example, the neurotransmission (4).

Thus, despite fantastic progress in molecular and cellular biology, the current concept of NES is frequently being revised, particularly in regard to the interface between the neuroendocrinology and the immune-modulatory systems.

Limited proteolysis processes leading to neuropeptides and peptide hormones distinguish themselves from the processes previously discussed in this book, because the neuropeptides and peptide hormones are produced from a very selective class of secretory proproteins which are hydrolyzed at specific peptide bonds by exquisite processing enzymes. These proteins (proproteins and processing enzymes) are enriched in the *trans*-Golgi network and the immature and mature secretory granules of the neuroendocrine cell system, such as the β-cells of the pancreas [146], the neuro- and adeno-hypophysis [147], brain, and other cell types that produce neuroendocrine peptides [143]. The mature bioactive peptides are usually stored in intracellular granules and are secreted into the interstitial space, including but not limited to the synapse junctions, where they interact with local receptors. They can also act in the distance as hormones, after being secreted into the bloodstream and reaching their receptors at target organs (Figure 7).

As previously presented, the study of the physiopathology of human diseases has been an important motivation for studying the roles of bioactive peptides in animal homeostasis, and survival in the environment. Food allergy, the damage caused to the GIS by the celiac disease, arterial hypertension, and the damage caused by snake bites to the CVS are some of the examples of human pathologies that impelled scientists to discover bioactive peptides produced by limited proteolysis. Studies on the physiopathology of human diabetes, particularly the work carried out by Donald Steiner and coworkers, provided the conceptual framework of the biosynthetic pathways that lead to the production of neuropeptides and peptide-hormones. The biosynthesis of insulin is a good example of how small steps, made during more than 20 years, can lead to the mainstream of neuropeptides biosynthesis [146].

Since the discovery of insulin in 1921 by Frederick Banting and Charles Best [148], a number of contributions have helped reveal the biosynthetic pathway that leads to the mature hormone. Indubitably, the determination of the structure of insulin by Frederick Sanger [149] represents not only the landmark of protein structure determination but it also was the starting point for the studies on insulin biosynthesis. This small protein (MW 5,808 Da, 54 amino acid residues), consisting of two peptide chains (A and B), bound to each other by two disulfide bridges, represented an immense challenge for Steiner and his coworkers, who aimed at understanding how Langerhans cells synthesized and released this hormone into the circulation.

Many fundamental scientific and technological developments occurred during the 1960s and 1970s and some laboratories contributed decisively to establish the basis of the contemporary cell and molecular neuroendocrinology: the mechanism of maturation and secretion of enzymes from the exocrine pancreas (George Palade), the importance of the removal of the signal peptide for the transport of nascent protein molecules across endoplasmic reticulum membranes (Günther Blobel), the structural basis of protein folding (Christian Anfinsen), the importance of the amino acid sequence containing a susceptible peptide bond, in order to be able to understand the specificity of proteolytic enzymes (Israel Schechter), the development of solid-phase peptide synthesis (Bruce Merrifield), the discovery of the processing enzyme that promotes the α-mating factor secretion in *Saccharomyces cerevisiae* (Jeremy Thorner), are certainly among those, who helped to understand insulin biosynthesis, later extended to the biosynthesis of other peptide hormones and neuropeptides.

The structure of insulin represented a fantastic challenge for the understanding of the biochemical pathway leading to the mature hormone. In contrast to most peptide hormones and neuropeptides, insulin is a large bioactive peptide, exhibiting stable secondary and tertiary structure. The fundamental question that had to be answered was whether the mature insulin molecule is generated by establishing two disulfide bridges between two independent chains (A and B) of peptides, as suggested by Humbel [150], or if insulin resulted from a single precursor protein to be proteolytically processed, as suggested by Steiner and coworkers. The answer to this simple but fundamental question was partially given, when Steiner and co-workers isolated an inactive protein (pro-insulin) from human adenoma of Langerhans cells (insulinoma) which, after *in vitro* treatment with pancreatic trypsin and carboxypeptidase B, gave rise to mature insulin [151, 152, 153, 154]. Interestingly, similar to the discovery of bradykinin [14], trypsin helped to demonstrate that insulin was a product of proteolytic digestion of a precursor protein.

Pro-insulin was finally isolated and sequenced by Ronald Chance and colleagues at Eli Lilly Laboratories [155]. However, this sequence did not allow the assumption, that a proteinase would be able to release the mature insulin from its precursor, unless the correct folding of the precursor could be presented for proteolytic processing. The predictions of the structural requirements for the correct processing of bioactive precursor proteins are currently subject of intensive studies [156, 157].

How and where should this processing occur in the Langherhans cells? Two fundamental contributions should be mentioned for the understanding of the mechanism of protein secretion: 1) the famous studies of Jamilson and Palade on the secretion of proteins from the exocrine pancreas [158] and 2) the discovery that nascent proteins are transferred across membranes of the endoplasmic reticulum [159]. Parts of these studies were used by Steiner and his collaborators to show, by pulse-chase experiments, that pro-insulin accumulates within rat pancreatic islets, and then undergoes conversion to insulin inside the cells. In a process requiring energy, the traffic of the newly synthesized pro-insulin molecules from the endoplasmic reticulum to the Golgi region was found to take about 15–20 minutes. There, the conversion to insulin takes place according to a pseudo-first-order kinetics, with a half-life of 30–60 minutes. Before insulin is secreted in response to glucose stimulation, the molecules are stored in mature granules of the β-cells. The whole process takes approximately 90 minutes [160]. Finally, specialized proteolytic processing occurs within the newly formed secretory granules [161].

Summarizing, today we know that the hormone derives from a single-chain precursor, pre-proinsulin, which is converted to proinsulin in the process of translocation into the endoplasmic reticulum [162]. The removal of the remaining basic amino acid at the C-terminus of the peptides by a carboxypeptidase B-like activity is required to release the mature insulin. In the late 1970s, the carboxypeptidase that produced insulin was reported to be a metalloenzyme with an acidic pH optimum, but was not purified or further characterized at the time [152]. In 1982, the carboxypeptidase that produced enkephalin was identified in adrenal chromaffin granules [163]. This enzyme was subsequently found to produce insulin in the pancreas [154].

Despite the progress on the peptide-producing carboxypeptidase, studies on the endopeptidases involved in neuroendocrine peptide production remained elusive. In the mid 1980s, John Hutton and colleagues identified two enzymes that cleaved proinsulin at the C-terminal basic residues, producing the intermediate that was the substrate for the carboxypeptidase. These proinsulin-processing enzymes were named type 1 and type 2 [164, 165]. Although identified, the proinsulin-processing enzymes were difficult to purify and sequence, and so their molecular identity remained uncertain. Meanwhile, working in yeast, Jeremy Thorner and colleagues identified a calcium-dependent peptidase, Kex2p, that cleaved the α-mating factor precursor (pro-α-mating factor) after the pair of basic amino acids [166]. In the early 1990s, cDNA sequences encoding two kex2p-like protein were isolated and sequenced; the resulting proteins were named prohormone convertase 1 (also known as prohormone convertase 3 and commonly referred to as PC1/3) and prohormone convertase 2 (PC2) 167; [see 168 and 169]. Soon after their discovery, the pro-insulin-converting enzymes originally identified by Hutton and colleagues were found to correspond to PC1/3 and PC2 [170].

These enzymes were shown to be involved in the processing of the peptide hormone precursor, proopiomelanocortin (POMC). Actually, PC2 and PC1/3 are considered to be the major prohormone convertase family members, involved in the processing of a large number of prohormones and neuropeptides, since they are highly expressed in the neuroendocrine system.

Limited proteolysis of proproteins at specific sites, generating multiple bioactive products, is one of the attributes of this class of proteinases. They share specificity preference for paired or single basic amino acid residues within the pro-neuropeptide and prohormone precursors, the major focus for the action of the family of prohormone-convertases. The reason for the existence of such a preferential cleavage site could be the higher accessibility of these hydrophilic points of the substrate, while the remaining parts of the precursor protein might be bound to the membrane of the cell [171]. The hydrolysis of the remaining basic amino acid at the C-terminus of the bioactive peptide is performed by carboxypeptidase E (CPE), a carboxypeptidase B-like enzyme, thereby completing the maturation of many bioactive peptides [163]. In some cases, peptides require additional modifications for biological activity, such as C-terminal amidation [172].

At present, the number of prohormone convertases stands at seven: furin and PC7 (the most ubiquitous mammalian convertases), PC2, PC1/3, PC4, PACE4, and PC5/-6A. A number of them are involved in constitutive processing events, furin and PC7, for instance [173], implicated in homeostasis, while PC5/-6A and PACE4 are involved in the activation of the (TGF)-β-like precursor [174]. The dibasic residues Lys–Arg and Arg–Arg most often flank neuropeptide sequences within the precursor, however, the dibasic sites Lys–Lys, and Arg–Lys, also occur [see 169].

The well-known prohormone and neuropeptide convertases PC2 and PC1/3 act on the proopiomelanocortin (POMC; precursor of ACTH and β-endorphin), proenkephalin, prodynorphin, prosomatostatin, proinsulin, promelanin concentrating hormone, secretogranin II, chromogranins A and B, proPACAP, and the neurotrophic factor proBDNF. There are more than one hundred different neuropeptides and peptide hormones, and new ones are still being discovered. The selective expression of PC2 and PC1/3 in neuroendocrine cells suggests the importance of these components of the PC enzyme family as prohomone processing proteinases [see 169].

As previously described for the RAS and the KKS, limited proteolysis of distinct families of pro-neuropeptides and pro-peptide hormones frequently generate a variety of bioactive peptides, able to bind to a number of distinct receptors, expressed by different tissues or organs, eliciting a multitude of biological activities. One of the best examples of this complexity is evidenced by the opioid peptides. Enkephalins, dynorphins, and β-endorphin are released from their respective protein precursors (preproenkephalin, preprodynorphin, and proopiomelanocortin, respectively) by the PC2 and PC1/3 convertases [see 169]. All opioid peptides share a common N-terminal Tyr–Gly–Gly–Phe signature sequence, which interacts with the mu, delta, and kappa opioid receptors to elicit their opioid and non-opioid activities. These receptors belong to the large GPCR superfamily, broadly expressed throughout the peripheral and central nervous systems. The opioid peptides Leu- or Met-enkephalin, derived from pro-enkephalin and pro-dynorphin, which are extended at the C-terminus (ligand of the kappa-opioid receptor), can be converted into Leu- and Met-enkephalin (delta ligand of opioid receptor) by oligopeptidases, such as EOPA and thimet-oligopeptidase, found in brain and peripheral tissues [175, 176, 177].

In the CNS, the opioid peptides play a key role in modulating mood and well-being, as well as addictive behaviors (hedonic control) [178]. They are also involved in peripheral actions, such as in the GIS, where they were suggested to regulate the secretion and expression of gastrointestinal mucins through activation of the μopioid pathway [34] (Figure 8).

The physiological importance of PC2, PC1/3, and CPE was deduced from studies on defective expression of these enzymes, or in gene knockout models affecting the biosynthesis of specific bioactive peptides. For instance, the absence of CPE causes the accumulation of intermediate peptides in the Langerhans cells. This defect may be responsible for obesity [153]. PC1/3, PC2, and CPE have been reviewed elsewhere [169, 179, 180].

It is important to mention that the selectivity of limited proteolysis is vital for animal survival, be it a specific animal or different animal species. It affects several families of neuropeptides and peptide hormones, for instance in vertebrates, the bioactive products of POMC, expressed in epithelial cells of the pituitary gland, differ according to their localization: the anterior part (corticotropes) produces predominantly ACTH and β-LPH upon the action of the PC2 proteinase, whereas the intermediate part (melanotropes) produces α-MSH, γ-MSH, CLIP, β-MSH, β-endorphin upon the actions of PC2 and PC1/3.

On the other hand, in different animal species, the products of POMC are highly influenced by the physiological needs of the animal. For some amphibian and reptiles, for instance, the production of α-MSH in the intermediate pituitary is essential in order to change the color of the skin, and for feeding behavior [181, 182].

Is the protein convertase family, coded by the *Pcsk1* to *Pcsk9* genes, the only enzyme family responsible for limited proteolysis of neuropeptides and peptide hormones?

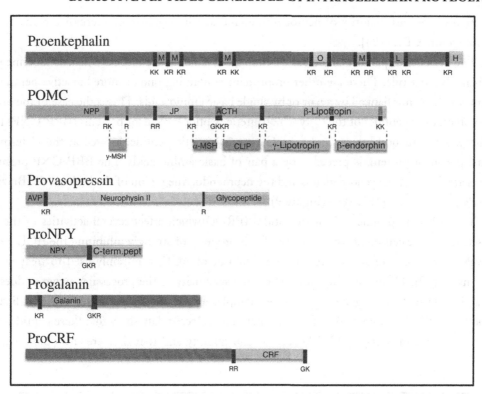

FIGURE 8: Proteolytic processing of proneuropeptides. Proteolytic processing is required to release the active neuropeptide from its proneuropeptide precursor. The proenkephalin contains multiple copies of the active neuropeptide: four copies of (Met)enkephalin (M), one copy of (Leu)enkephalin (L), and the related opioid peptides ME–Arg–Phe (H), and ME–Arg–Gly–Cleu (O). The proopiomelanocortin (POMC) contains different active peptides within its structure, originating the distinct peptide hormones ACTH, α-MSH, and β-endorphin. Some precursor proteins like proNPY, progalalin, proCRF, and provasopressin contain one copy of the active neuropeptide. Lysine (K) and/or arginine (R) represent monobasic, dibasic or multibasic sites where proteolytic processing occurs. Figure from reference [169].

Recently, a cysteine protease pathway was suggested to function in proneuropeptide processing. Cathepsin L is able to cleave at a dibasic pair of amino acids as well as at a monobasic site and was proposed to be able to process proenkephalin as well as other proneuropeptides. This enzyme cleaves to the N-terminal side of basic residues, and therefore required the action of an Arg/Lys specific aminopeptidase to produce mature peptides [183]. However, it is unlikely that this pathway functions in neuropeptide processing as proposed based on the finding that mice lacking CPE activity have extremely low levels of most neuropeptides, indicating that CPE is involved in the production of the vast majority of neuropeptides and arguing against an alternative pathway that does not require CPE [184].

The presence of pairs or single basic amino acids flanking the bioactive neuroendocrine peptides is not a strict rule. There are other proproteins containing one or more bioactive peptides in tandem which are not flanked by a pair or by single basic amino acids. This is the case of the bradykinin-potentiating peptides of the C-type natriuretic peptide-precursor proteins (BPP-CNP) from *Bothrops jararaca* tissues. Interestingly, the C-type natriuretic peptide, located at the C-terminus of same precursor protein, is preceded by a pair of basic amino acids. The BBP-CNP precursor proteins are found in the venom glands and the neuroendocrine region of the brain of the Brazilian snake *Bothrops jararaca* [185, 186] (Figure 9).

The BPPs are proline-rich oligopeptides (PROs) which affect several activities of the cardiovascular and the central nervous systems. Their property of strongly inhibiting the ACE helped to develop Captopril, an active-site directed inhibitor of ACE, commonly used to treat human hypertension [92]. There are other cases where the specificity of the processing enzymes does not seem to be adjusted to the general rule. Few examples are: 1) the kisspeptins are GPCR ligands, originally identified as suppressors of human metastasis. Recent data show that these peptides have an important role in initiating GnRH secretion in puberty and that they are also opioid peptide

-VQQA**QSWPGNIPP**LK-	**BPP1**
-VQQWA**QGGWWPRPGPIPP**LT-	**BPP2**
-VQQWA**QNWPHPQIPP**LT-	**BPP3**
-VQQWA**QGRAPGPPIPP**LT-	**BPP4**
-VQQWA**QGRAPHPPIPP**APL-	**BPP5**

QKWAPL**QKWAP**LLQPH..........LKGLAKK**GAAKGCFGLKLDRIGTMSGLGC**

BPP6 **BPP7** **C-Type Natriuretic peptide**

FIGURE 9: Different sites for proteolytic cleavage. Sequences of seven Bk-potentiating peptides (BPP 1 to 7, bold, underlined sequences) and of the C-type natriuretic peptide (bold sequence at the C-terminus of the proprotein), deduced from the cDNA clone of the precursor proprotein from the *Bothrops jararaca* venom gland. Except for the sequence of the C-type natriuretic peptide, which is preceded by a pair of basic amino acids (KK, in red), all other bioactive peptides have no basic amino acid flanking their sequences.

regulators [187, 188]; 2) the sarafotoxins, endothelin-related peptides of the venom glands of the snake *Atractapis enganddensis* [189]. These toxins also affect the cardiovascular system of the prey, using a similar mechanism as endothelin. The processing enzymes responsible for the release of these bioactive peptides have not yet been identified.

Finally, it is worth mentioning that evolutionary studies are revealing neuropeptides and peptide hormones in vertebrates and invertebrates, seemingly, an inexhaustible source of neuropeptides. Some of them derived from ancestral neuropeptide precursors of known mammalian peptide hormones or neuropeptides. Among many other examples, the vasopressin/oxytocin/neurophysin prohormones [190] have been extensively studied. They gave rise to the discovery of a novel family of neuropeptides, found within the neurophysin sequence: in the sea urchin, the myoactive neuropeptide, NGFFF-amide, is structurally unrelated to the oxytocin/vasopressin-type neuropeptides [191], and the neurophysin peptide of Branchiostoma, the SFRGV-amide was described as sharing 100% similarity with the N-portion of the mammalian neuropeptide S involved in arousal and anxiety [192].

In summary, by briefly reviewing the limited proteolysis processes of neuropeptide and peptide hormone precursors, we intended to stress their role in releasing peptide hormones at different time points and locations of the animal body during its life span, thus contributing to survival and reproduction. We also intended to give an idea of how far we are from fully understanding the mechanisms and circumstances of the release of neuropeptides and peptide hormones, as well as from the whole spectrum of their actions.

3.3   PEPTIDES FROM MITOCHONDRIA

As opposed to the crucial need for genome stability, the global and selective degradation of proteins is critical for the survival and well-being of living organisms [2]. It is represented by the fine-tuning of biological networks in order to respond to small changes in protein concentrations, as well as by mechanisms for protein folding, for corrections, and/or clearance, which are part of the evolutionary history of the cell. These aspects, which have not yet been sufficiently described, are demanding special attention nowadays, since the protein stabilization index (PSI) is crucial in most cellular processes, including cell cycle progression, signal transduction, and differentiation [89, 193]. The articles mentioned above, including their bibliography, highlight the obvious importance of this subject for basic science, mainly in regard to biological regulation, causing an impact on studies related to body development and maintenance, suggesting effective strategies to treat degenerative diseases, cancer, virus infections, among others.

Bioactive peptides have recently been shown to play important roles in proteostasis. Maintenance of protein stability in mitochondria illustrates the importance of bioactive peptides. Young et al. [194] demonstrated in *Saccharomyces cerevisiae* that peptides between 600 and 2100 Daltons, generated by mitochondrial proteolysis, are actively transported to the mitochondrial intermembrane space through ABC-transporters, reaching the cytosol by passive diffusion. Thereby, peptides allow the communication between mitochondria and cellular environment [195].

Later, Haynes et al. [196] demonstrated by genetic analysis in *Caenorhabditis elegans* that stress caused by misfolding of mitochondrial matrix proteins signals to the nucleus of the cell through the bZIP protein to express proteins to correct the perturbed folding. Thereby, the organelle reacts against loss of thermodynamic stability and aggregation propensity of unfolded proteins by eliciting a response that includes the expression of the nuclearly encoded protein, the ubiquitin-like ClpXP, that localize to mitochondria. Its activation is performed by the homeobox containing transcripton factor bZIP, which in turn is suggested to be activated through different steps involving peptides generated by the proteolytic components of Clp-proteases, which are released into the cytosol by an ABC-transporter. Together, the expression of these proteins represent the cell response to damage produced by irreversible aggregates and unfolded proteins affecting cell functions and survival. Generally speaking, this response is elicited to maintain protein homeostasis (proteostasis) by increasing the expression of heat-shock proteins (chaperones) to assist the refolding process, the proteolytic systems in charge of degradation of unfolded proteins, thus, representing an efficient signal to the cell genome, mediated by bioactive peptides [196] (Figure 10).

Protein misfolding and aggregation have a significant impact on cell fitness. An increasing number of degenerative disorders, including Alzheimer's disease and diabetes, amyloid fibrils in humans, and neurotoxicity have been linked to these dysfunctions [12].

3.4   PEPTIDES TRANSLATED BY RIBOSOMES

Oligopeptides are beginning to unravel the route to answer fundamental questions in developmental biology by demonstrating their roles in cellular organization of embryos. The *tal* gene of *Drosophila melanogaster* expresses a 1.5-kilobase (kb) transcript containing several open reading frames (ORFs), which code for peptides smaller than 50 amino acids [197]. Surprisingly, peptides translated from these ORFs are just 11 amino acids in length and mediate gene function in the fly's development. In contrast to bioactive peptides produced by limited proteolysis of precursor proteins, these oligopeptides are directly generated from transcripts (sORFs, <100 codons). They represent the smallest gene products known to date. Functional analysis of this gene in the fruit fly shows that it has important functions throughout development, including tissue morphogenesis and pattern formation [198, 199].

During Drosophila embryogenesis, peptides ranging from 11 to 32 amino acid residues are involved in cell differentiation and body development [198, 199]. These peptides act by transforming a repressor protein into an activator of the transcription factor, thus providing a strict temporal control to the transcriptional program of epidermal and tracheal morphogenesis. They promote the

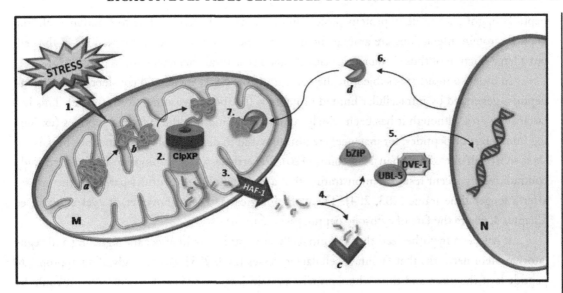

FIGURE 10: Mitochondrial peptides signaling. Upon mitochondrial stress in *C. elegans* (1), some proteins (a) get a misfolded structure (b) and are degraded by the proteinase complex ClpXP (2), generating peptides. These peptides are exported through the ABC transporter HAF-1(3), thus transmitting mitochondrial stress signals to the cytosol. The peptide binds to a specific receptor (c), which activates the transcription factors bZIP and UBL5/DVE1 (4). Once translocated to the nucleus (5), they activate the expression of mitochondrial chaperones (d) that migrate to mitochondria (6) and assist in the refolding process (7). M: mitochondria; N: nucleus.

release of the activator by assisting at the removal of a fragment of about 50 amino acid residues from the N-terminus of the repressor protein [198, 199]. These experiments strongly suggest that peptides probably regulate additional developmental factors. In fact, recent studies indicate the existence of thousands of unexplored polypeptides of less than 100 amino acids, originated by similar mechanisms in mice and humans [199, 200, 201]. Future functional analyses should elucidate how these small peptides contribute to various biological processes, including development and differentiation.

3.5   PEPTIDES AND PROTEIN–PROTEIN INTERACTIONS

Except for the peptides described above, which derive from mitochondria or are translated directly by the ribosome, the other bioactive peptides, once released, act on "classical" receptors (GPCRs, MHCs, for example), eliciting their biological effects. However, the roles of bioactive peptides are in the imminence of being substantially expanded to other areas, especially to those, in which peptides might play a role in protein-protein interactions. In fact, according to London et al. "the peptide–protein interactions are among the most prevalent and important interactions in the cell, but a large fraction of those interactions lack detailed structural characterization" [202].

In order to insert this subject in the context of this book, we should consider the fate of all peptides generated by intracellular limited proteolysis (by the proteasomes, for instance). This is a fascinating area, although it has been widely accepted that most of the cytosolic peptides (excluding immunogenic peptides, for instance) are not supposed to display biological functions because of their short half-life of less than 10 seconds [120]. Nevertheless, this assumption has been strongly contradicted by recent results demonstrating that a large number of peptides persist in the cytosol after a longer time frame [203, 204]. Their physiological activities, however, are yet elusive (see Chapter 4, where the fate of cytosolic peptides is discussed).

It has been hypothesized that the intracellular peptides could affect the signaling and regulatory protein networks that facilitate cellular processes [203, 205]. Accordingly, if an appropriate peptide is at the place and time where protein–protein interactions are likely to occur, it could be recruited to interact with the interface, thus accomplishing a specific role in such an interaction [see 206].

The androgen receptor, which belongs to a subfamily of nuclear steroid receptors, may be a good example to illustrate the importance of this issue. Receptor activation by androgens regulates male sexual differentiation, development, and maintenance of male reproductive tissues. Androgens induce a conformational change in the ligand binding domain, which results in the formation of a novel hydrophobic co-activator receptor, whose binding motif is the sequence LXXLF [207]. Peptides containing this binding motif may elicit transcription initiation after binding to this androgen co-activator receptor.

A similar rational could be used to explain the mechanism of action of bradykinin-potentiating peptides (BPPs) from the snake *Bothrops jararaca* venom. These peptides belong to a family of proline-rich oligopeptides (PROs), which share the sequence motif PXIPP at their carboxy-termini. Their biological activity, ultimately, causes a strong reduction of the systemic blood pressure in hypertensive animals (PROs are not hypotensive in normotensive animals) due to synergistic actions of these peptides, which include inhibition of ACE, activation of arginino-succinate synthase and yet unidentified targets that increase the release of nitric oxide in the CVS [see ref. 89]. The sequence motif of the PROs is likely to affect regulatory proteins, such as SH3 of tyrosine kinases, WW, HLA-DQ2, and PDZ proteins [50, 208].

Peptides, especially those linear peptides ranging between 5 and 15 amino acid residues, present several advantages, when compared to larger polypeptides or to their protein competitor counterparts. Their flexibility, the presence of two or three hot spots (frequently amino acid residues W, L, F, Y, I) within the internal amino acid sequence, as well as the backbone hydrogen bond donors or acceptors, make them more amenable than proteins or larger peptides to accomplish the interactions with the partner, causing very small conformational changes at the protein interface.

Besides the importance of the physiological participation of bioactive peptides in processes involving protein–protein interactions, the understanding of peptide–protein interactions, in general, is certainly necessary to develop new and more specific inhibitors to be used in several pathologies [see 206].

. . . .

CHAPTER 4

Proteolytic Enzymes

We have already pointed out that the most frequent proteolytic enzymes that generate bioactive peptides are tightly controlled both in the extracellular (see RAS and KKS) and in the intracellular media (see lysosome/endosome, UPS, and NES). In addition, other proteinases generate bioactive peptides, either by chance (like the glutamine/proline-rich peptides derived from partial proteolysis from gluten proteins in the GIT) or by not well-known proteolytic processes, as for example, promoted by caspases 5 and 10 [209], nardilysin [210], or the insulin-degrading enzyme [211].

In general, the process of bioactive peptide generation is accomplished by the synergic actions of one proteinase and one or more peptidases. Its efficacy relies on the specificities of the proteolytic enzymes toward their substrates. Intriguingly, a limited number of proteolytic enzymes, approximately 700, is encoded in the genomes of plants and animals [212]. Their proteolytic action is directed toward countless substrates harboring putative bioactive peptides. The structures of proteolytic enzymes from both, plants and animals, are compatible with one of six "classic" catalytic types: aspartic, cysteine, glutamic, metallo, serine, and threonine peptidases [213]. Consequently, it seems reasonable to hypothesize that the molecular evolution of proteolytic enzymes and their substrates is coupled, thus allowing limited proteolysis to release specific bioactive peptides for biological tasks, whenever they are needed, in order to maintain homeostasis.

Examining the substrate structure and the specificity of proteolytic enzymes which lead to bioactive peptides, a separate analysis of proteinases and peptidases is required (nomenclature of Fruton [11]). In fact, the two classes of proteolytic enzymes have drastically different structural requirements: the proteinases recognize the susceptible peptide bond within more rigid protein-substrates, either folded or unfolded, whereas the peptidases make such recognition within more flexible peptide-substrates.

4.1 THE PROTEINASES

How do proteinases excise bioactive peptides from native proteins? For food digestion, autolysis, and energy-dependant proteolysis, the physical chemistry conditions of the medium (pH of the stomach and the lysosome), or the chemical modification of the protein (UPS and ClpX-proteinases), drastically modify their 3D structure, increasing accessibility to the susceptible peptide bonds for the proteolytic attack. Thus, it is expected that a larger number of peptides is generated, as compared to the one obtained by proteolysis of folded protein-substrates in the RAS, the KKS, or the NES, for example. In the first case, the probability to obtain a bioactive peptide by chance or by unknown mechanisms is higher than in the latter case.

For folded proteins at physiological conditions (neuropeptide secretory pathway, for instance), the answer to how proteinases excise bioactive peptides from precursors is ultimately related to the structural requirements for the enzyme/substrate interaction, such as to lead the susceptible peptide bond of the substrate to reach the catalytic center of the proteinase. More specifically, the susceptibility of a folded protein to limited proteolysis is determined by: (i) co-localization of the protease and the substrate in both space and time, fulfilling *in vivo* requirements; (ii) the presence of an amino acid motif (a sequence of few amino acids) containing the susceptible peptide bond, adjustable to the enzymatic cleft; (iii) the spatial dimension that accommodates the amino acid motif to the binding sites; (iv) a possible need for adjusting fit between the enzyme and the substrate, which may be performed by the interaction of an activator at a specific site, away from the active site (exosite) [157].

Today, we know that the 3D structural context of the amino acid sequence containing the cleavage site [nick-site] [156], or P1 (according to the nomenclature of Schechter and Berger [214]), seems to play the most prominent roles [215]. Typically, limited proteolytic sites are found at flexible loop regions, which are exposed to the solvent. The latter condition is notably absent in regions of regular secondary structure, especially in β-sheets.

According to Hubbard [156] the portion of the molecule containing the nick-site undergoes a conformational change, allowing the penetration of, at least, 10 or more amino acid residues into the groove, prior to cleavage. The same authors developed an algorithm, containing several conformational parameters that could successfully predict the sites, where limited proteolysis should occur within a protein substrate. Although Hubbard et al. [156] used a small number of proteinases, known to be involved in limited proteolysis, they were able to provide the conceptual framework of the structural features that regulate limited proteolysis [157].

Our current knowledge of mechanisms allowing the prediction of the structural features that trigger the proteolytic attack on a protein-substrate is crucial for the understanding on how bioactive peptides are released from their precursors. The conformational parameters that allow determining the mechanism of proteolytic processing with precision are far from being understood. Even for proteinases showing strict primary sequence specificity (as for trypsin, thrombin, PC2, and PC1/3, for instance), it is not always easy to be completely sure which site, among many putative ones, will be preferentially hydrolyzed. For proteinases exhibiting broad specificity (subtilysin, elastase, termolysin), the prediction is even more challenging, especially for those requiring hydrophobic residues at P1. A number of studies indicate that accessibility, rather than flexibility, is considered an essential structural determinant for limited proteolysis [156, 216]. At physiological conditions, a number of variables, difficult or impossible to be reproduced *in vitro* conditions, is likely to influence the accessibility of the susceptible peptide bond to the catalytic site of the enzyme. As pointed out by Lu et al. [171], the reason for the existence of such a preferential cleavage site could be the higher accessibility of these hydrophilic points of the substrate (nick sites), while the remaining parts of the precursor protein might be bound to other cellular structures (i.e., proteins) [171]. Examples of this situation are described in this book (Chapters 2 and 3).

Serine-proteinases offer the best-known model of peptide bond hydrolysis in the processes of limited proteolysis [217]. About 1/3 of the proteolytic enzymes are serine proteinases, evolutionary derived from trypsin. These enzymes take part in fundamental processes of animal homeostasis, such as in food digestion, blood coagulation, bioactive peptides generation, fibrinolysis, complement activation, fertility, apoptosis, and immunity.

4.2   THE OLIGOPEPTIDASES

Between 1938 [11] until the late 1970s, the known peptidases were the exopeptidases (carboxy-, amino-, di-, tri-peptidases [218, 219]), and the proteinases were the endopeptidases. Together, these enzymes are mostly involved in general proteolysis.

Increasing medical interest for peptides acting on specific receptors and regulating essential physiological processes, and the analogy with the metabolism of non-peptide signaling molecules (acetylcholine, cathecolamines, etc.), led pharmacologists to speculate, whether specific peptidases were in charge of the conversion and inactivation of bioactive peptides. This question was particularly important in order to evaluate the putative role of peptidases in the CVS in regard to the metabolisms of angiotensin and bradykinin (Chapter 2). The arterial hypertension could result from dysfunctions of peptidases, especially those of the endothelial cytoplasmic membranes, involved in the conversion and inactivation of A-II and Bk, such as the ACE.

In the early 1970s, two metallo-ectopeptidases were discovered: ACE (EC 3.4.15.1), a carboxy-dipeptidase of the plasma membrane of endothelial cells, and NEP (EC 3.4.24.11), located in the brush-border of the kidneys [220, 221]. As described in Chapter 2, these two enzymes, in particular ACE, caused an enormous impact on the biomedical sciences because it carries out the conversion of A-I to A-II [88], as well as the inactivation of bradykinin [222, 223]. Studies on ACE and its inhibitors led to the development of captopril, the first active center-directed inhibitor of the enzyme, revolutionizing the treatment of human hypertension (see [89, 224]).

Coincidentally, a distinct sub-class of tissue endopeptidases was suggested to explain the peculiar activity of not yet known peptidases from rabbit brain. In contrast to the proteinases, they were endopeptidases that selectively hydrolyzed internal peptide bonds of oligopeptides. These findings resulted from the study of the short lasting, but dramatically strong effects of Bk on the CNS of rabbits. The short duration of the effects, elicited by Bk, were hypothesized to be due to Bk's rapid inactivation by brain peptidases [225]. The search for the enzyme(s) responsible for Bk inactivation led to the discovery of two thiol-activated endopeptidases in rabbit brain cytosol [226, 227].

Since the activities of these enzymes were restricted to the hydrolysis of a specific bond within oligopeptide sequences [55], they were designated endo-oligopeptidase A (EOPA) and endo-oligopeptidase B (EOPB). They present different tissue distributions: EOPA is present in the cytosol of the CNS and in neuro-endocrine tissues [228, 229, 230], whereas EOPB is ubiquitously distributed [56]. During the 1980s until the early 1990s, EOPA and EOPB were shown to be able to hydrolyze a number of bioactive peptides, such as Bk, A-I and II, LHRH, neurotensin, opiod peptides, and several derived molecules thereof. The overall results allowed to conclude that: i) the peptidase activity was restricted to peptides of 8–13 amino acid residues in length; ii) EOPA did not present a clear preference for residues at P1, whereas EOPB is a post-proline cleaving enzyme [176, 231, 232, 233, 234, 235, 236].

The complete molecular characterization, achieved in early 2000, revealed that EOPA (EC 3.4.22.19) is a 38-kDa protein, which maps to the human chromosome 17p13.1 [237]. It was characterized as a cysteine-peptidase [238], whose reactive cysteine (Cys^{273}) has been identified [239]. Surprisingly, at the end of 2000, EOPA was rediscovered, not as a peptidase, but as the nuclear-distribution gene E homolog like-1 product (Ndel1) [240, 241].

EOPB was simultaneously discovered by Oliveira et al. [227] in the cytosol of rabbit brain, and by Walter in rat kidney [242]. EOPB is a typical post-proline cleaving serine peptidase [243, 244]. Its further characterization led it to be referred to as prolyl oligopeptidase or POP (EC 3.4.21.26). POP is a member of a serine-peptidase family comprising the dipeptidyl-peptidase (DPP-IV, EC 3.4.14.5), the acylpeptide hydrolase (APEH, EC 3.4.19.1), and a protein lacking catalytic activity [213; see also 230]. A peculiarity of POP is its thiol-activated feature, which was first recognized at its discovery, being later extended to the other aminopeptidase family members [212, 227].

In 1983, two other cytosolic metallo-oligopeptidases were discovered: thimet oligopeptidase or TOP, EC 3.4.24.15 [246], and neurolysin or NL, EC 3.4.24.16 [247]. Together EOPA, POP, TOP, and NL were considered to belong to a sub-class of tissue peptidases, and are named oligo-peptidases [219].

The full characterization of TOP and NL showed that they belong to the M3 family of metallopeptidases, displaying Zn^{++} at the catalytic site (sequence motif HEXXH), and sharing 60% sequence identity among them [248]. Interestingly, EOPA, TOP, and NL exhibit very similar specificities toward the same neuropeptides [235, 236, 249, 250], thus suggesting a similar recognition mechanism for the susceptible peptide bonds.

Not only the cytosolic peptidases (EOPA, POP, TOP, and NL) but also the extracellular peptidases (ACE and NEP) exhibit selectivity for oligopeptides [207, 208], classifying them as candidates to be included into the sub-class of proteolytic enzymes designated oligopeptidases. Crystal structures of these enzymes support the strict selectivity for oligopeptides [251, 252, 253, 254, 255]. The difficulty for an oligopeptide to penetrate into the groove, where the catalytic group of the oligopeptidase is located, can be illustrated using the example of POP. The peptide has first to penetrate a 4 Å hole on the surface of the enzyme in order to reach the 8,500 Å^3 internal cavity, where the active site is located. However, the presence of an internal proline within the oligopeptide sequence is not sufficient to guarantee that the peptide will be hydrolyzed. Interplay between ligand entry and egress of POP is likely to occur [245, 251, 256]. A similar peculiar structural feature may be extended to all oligopeptidases, thus allowing this sub-class of peptidyl-hydrolases to play specific physiological roles concerning their actions on bioactive peptides. Moreover, understanding the selectivity of these enzymes for specific features of oligopeptides is certainly important for the development of specific inhibitors. In fact, among other examples, it was shown that opioid peptides can either be substrates or strong competitive inhibitors of EOPA and TOP [235] (Figure 11).

The intracellular conversion and inactivation of secreted neuropeptides by oligopeptidases (Chapter 3), although possible, would require the penetration of the peptide into the cytoplasm of the target cell. Alternatively, the intracellular oligopeptidases could be externalized in order to carry out their proteolytic activity in the extracellular milieu [257]. If, on one hand, the physiological role of the extracellular peptidases in bioactive peptide metabolism (mainly ACE and NEP) has been firmly established [222, 223], on the other hand, the physiological role of the intracellular oligopeptidases is still elusive. More likely, it seems that the intracellular oligopeptidases (EOPA, POP, TOP, and NL) would participate in the intracellular peptide metabolism, since they are in the same milieu as their substrates, which are predominantly smaller than 20 amino acids in length. The strict selectivity for oligopeptides allows active oligopeptidases to be surrounded by functional proteins without offering any risk to their integrity. These peptidases could not only contribute to the final steps of protein turnover, but also complete the processing activities of proteinases, by trimming, converting, and/or inactivating bioactive peptides.

Most of the discussion on the action of peptidases on biosynthesis and degradation of bioactive peptides in this book is related to immunogenic peptides, mainly generated by the UPS (Chapter 3). In the following section, we will highlight the potential interplay between cytosolic peptides and peptidases in the complex milieu of the extra- and intracellular environments, which generate, destroy, preserve, and/or prepare peptides for biological actions.

Which is the fate of oligopeptides in the cytosol? If we accept the interpretation of Reits et al. [130], peptides are condemned to non-functionality since, according to these authors, peptides have

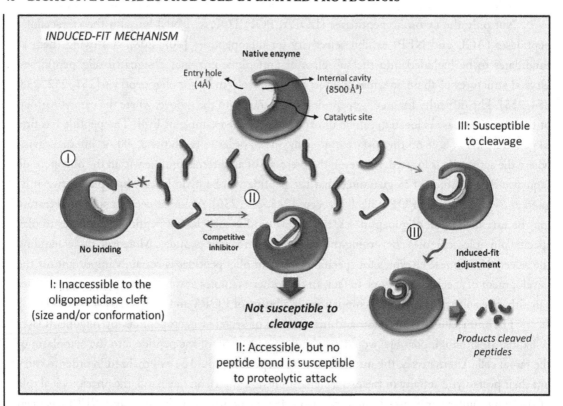

FIGURE 11: Induced-fit mechanism. Peptides exist in a variety of conformation stages that might influence their interaction with a specific enzyme. In its native form, the enzyme presents a narrow entry and a small internal cavity which contains the catalytic site. (III) The peptide (substrate) reaches the internal cavity of the enzyme and induces a large-scale conformational change (induced-fit adjustments), allowing the proteolytic attack. The peptide can be processed generating smaller fragments. (II) If the peptide gains access to the enzyme groove, but presents an inappropriate conformation, it may act as competitive inhibitor. (I) Other oligopeptides, unable to penetrate into the catalytic chamber, are neither substrate nor competitive inhibitors.

very short half lives ($t_{1/2}$ <10 seconds), and would be, therefore, inactivated before signaling their biological messages to "receptors." However, a number of peptides manage to be preserved to fulfill biological functions. What makes these peptides so special? Concerning the UPS, part of the answer may be found in post-proteasome modification. In a recent study, 69 out of 272 peptides found in the cytosol of three cell lines were acetylated, thus resistant to degradation [204]. The acylpeptide hydrolase (APEH), one of four members of the proline oligopeptidase family (see above), can remove N-acylated amino acids, allowing the action of aminopeptidases. Interestingly, the inhibition of APEH down-regulates UPS activity. Therefore, these recent results point to a promising target for the development of anti-cancer drugs, through down-regulation of UPS activity in malignant cells [258].

Other peptides are protected from degradation by binding to cytosolic proteins, such as the MHC-class I transport protein (TAP), or to oligopeptidases, such as TOP [259; and see below], to heat-shock proteins [260], or to not yet identified cytoplasmic molecules.

Concerning the role of oligopeptidases in bioactive peptide degradation, several questions should be addressed before attempting to introduce these enzymes into the pathway of intracellular bioactive peptides. One of the most important is: besides the peptide length, what is the base of enzyme specificity/selectivity?

Oligopeptides are flexible structures in aqueous solution, while, on the other hand, the oligopeptidases, such as, TOP and NL, display broad specificities. These two features of enzyme and substrate prompted us to assume that the catalysis has high probability to occur, no matter the amino acid sequence of the oligopeptide. This assumption, however, led many scientists to draw wrong conclusions, namely, that the specificity of oligopeptidases is not a relevant question to be discussed. For instance, in studies on MHC class I epitope degradation, the authors suggest that the activity of TOP is responsible for the decrease in antigen presentation [261]. Part of this conclusion is based on *in vitro* experiments, in which synthetic epitopes were shown to be degraded by recombinant TOP after over-digestion [262]. However, besides showing no consideration for TOP's specificity, it is likely that, at physiological conditions, TOP activity depends upon several environment variables, difficult or impossible to be reproduced in *in vitro* experimental designs. For example, the authors did not take into account that the presence of cytosolic substances is able to affect enzyme activity in the internal milieu of the cells. These factors include, for instance, the presence of putative competitive inhibitors, such as peptides, or ATP molecules [259, 263].

An array of experiments was carried out in several laboratories [see 250] to determine the specificity of oligopeptidases. Accordingly, enzymatic assays performed *in vitro*, using dozens of synthetic bioactive peptides of distinct amino acid sequences, revealed that different sequences or small changes in the amino acid sequences of a specific oligopeptide substrate can either modify the site of cleavage, or it can affect the K_{cat}/K_m ratio, leading to no interaction of the peptide with the enzyme, or, finally, turn the modified peptide into a competitive inhibitor [235, 236, 238, 259]. It is possible to conclude that the dynamic structure acquired by the peptide, while inside the catalytic cleft, determines whether or not a peptide bond can become susceptible to proteolytic attack; otherwise, it would be difficult to explain the number of proline-containing epitopes to survive the action of POP.

An attempt to approach the structural requirements, defining the specificity of recombinant TOP (rTOP) towards an oligopeptide, was performed by Jacchieri et al. [264]. According to this study, the chain conformation of TOP-substrates, in contrast to the chain conformation of non-substrates, are coiled, and exhibit sharp turns, coincident with the positions of hydrolysis by rTOP. These results suggest a possible rule for the specificity of TOP to hydrolyze a peptide bond within an oligopeptide. This rational could also be applied to explain the post-proline specificity of POP. In fact, it is well known that L-proline residues, within a peptide structure, limit the conformational freedom of the peptide [265, 266], and may expose the peptide bond involving the carboxy-side of the proline residue to proteolytic attack.

Thus, besides the peptide length, the conformation acquired by the peptide while accommodated within the oligopeptidase groove, seems to define the fate of the peptide, i.e., whether it is going to be hydrolyzed or preserved, in order to exert its biological function. Besides, it has been demonstrated that oligopeptidases, such as POP, are influenced by interactions of molecules with its allosteric site [256]. This kind of peptide interaction with a macromolecule reminds us of the property of a bioactive peptide binding to its receptor (see below "The non-peptidase role of oligopeptidases"). This hypothesis could represent a common structural feature of immunogenic peptides since it is also required for MHC class I peptides to be accommodated into the groove (cleft) of either the transporter associated with antigen presentations (TAPs) [267] or to the MHC-class I molecule itself [137, 268]. These features may not be just coincidental, but could rather be part of the cellular mechanism to assure peptide selection, integrity, and therefore, biological function.

Additionally, the physiological roles of oligopeptidases may include trimming inactive peptide precursors, leading to their active form [225], converting bioactive peptides into novel ones [236], inactivating them, in order to restrain the continuous activation of specific receptors [235], or protecting the newly generated bioactive peptide from further degradation, suggesting a peptide chaperon-like activity [259].

For the intracellular oligopeptidases, especially those involved in human diseases, the existing studies are promising but not yet as developed as for the ACE. TOP, a ubiquitous cytosolic oligopeptidase, is a remarkable example of how this enzyme could play an essential role in immune defense against cancer cells [210]. TOP has also been suggested to be involved in tuberculosis [269]. Other examples are: 1) the POP of nervous tissues has been suggested to be involved in neuropsychiatric disorders, like post-traumatic stress, depression, mania, nervous bulimia, anorexia, and schizophrenia [see 245]; 2) NEP has been implicated in cancer [270].

4.2.1   The Non-Peptidase Roles of the Oligopeptidases

Several oligopeptidases are thought to be involved in other processes which, apparently, are dissociated from the hydrolytic activity of the enzyme. For instance:

i) The ACE and the ACE2 exert a pivotal role in regulating angiotensin II levels. The oligopeptidase activity of these enzymes has focused most attention due to the medical importance of these enzymes for human physiology and pathologies. Recent discoveries, however, drove attention to unexpected roles of ACE, ACE2, and its homologue, collectrin, in intracellular trafficking and signaling. Activities, such as molecular chaperoning, cross talking with the B2 kinin receptor, a role as virus receptor, and in insulin secretion were attributed to these enzymes [see 271];

ii) The regulation and functioning of adult organisms is largely influenced by the micro-environment within the tissues. Disturbances of the fine balance of extracellular matrix synthesis and homeostasis may contribute to tumorigenesis. Protein degradation and structural organization (remodeling), occurring during morphogenesis or tumorigenesis, involve many enzymes which are important for cell survival, proliferation, migration, polarization, and differentiation [272]. NEP or CALLA is one of these enzymes, being ubiquitously present in the cytoplasmic membrane of the cells. NEP is now known as CD10-peptidase. This enzyme modulates the appropriate concentration of bioactive peptides in the cell environment, and is involved in a large number of physiological and pathological activities; including cell proliferation, tumor progression, and liver metastasis formation [see 270]. In addition to its enzymatic activity, involved in the regulation of a large number of processes mediated by bioactive peptides, CD10 is likely to function a as non-peptidase molecule. For instance, in neoplastic progression, CD10 could directly mediate signaling events, suggesting an indirect negative regulation of cell migration by CD10, independently of its catalytic function [273].

iii) As reviewed above, NL (as well as the NL-like metallo-peptidase) was first described as an ubiquitous cytosolic oligopeptidase that degrades neuropeptides, thus playing a putative role in a large number of activities, such as nociception, cardiovascular control, microvascular permeability, MHC-class I antigen presentation, etc. Besides presenting the typical oligopeptidase activity, NL (or NL-like oligopeptidase) was also identified in rat brain synaptic membranes [274], and in a number of other intra- and extracellular compartments [213, 257, 275]. However, the membrane-anchoring mechanism(s) and the differences among the membrane-bonds, as well as the mitochondrial, the nuclear, and the cytoplasmic variants are not fully understood. Recent studies revealed that NL is a non-AT1, non-AT2 angiotensin binding site in rat brain membranes [276].

iv) POP is present in all organisms including bacteria, fungi, plants and animals [277], suggesting its participation in a number of basic processes of living organisms. In mammals, POP is detected in different tissues, found at the highest level in the brain [278]. POP has also been shown to play a role in several physiological (brain development, for instance) and pathological processes (neurodegenerative disorders), [see 279], and in cell cycle control [280]. Although the enzyme inactivates several bioactive peptides, the mechanism by which POP exerts its biological effects remains unknown. Particularly interesting is the participation of oligopeptidases (EOPA and POP) in cell proliferation, and in the regulation of neurogenesis during brain development [239, 281]. It seems likely that, similar to EOPA, the role of POP in neurogenesis is not exclusively related to its peptidase activity [241].

v) In the nervous system, neuron migration during development brings different classes of neurons together in order to make appropriate interactions possible. A defect in this process during brain development in humans causes a serious malformation known as smooth brain or lissencephaly [281]. Several proteins converge to promote neuronal migration as well as other processes of neuronal development, including neurite outgrowth; one of them is NUDEL (Nuclear Distribution Element-Like), now known as Ndel1 (nuclear-distribution gene E homolog like-1). Surprisingly, the cDNA sequences of human Ndel1 and EOPA, which had been characterized earlier, showed them to be the same molecule (GenBank Acc. No. AY004871). Neurite outgrowth has been recently demonstrated to depend on EOPA's oligopeptidase activity. In fact, when Cys^{273} of EOPA is mutated to Ala, both the oligopeptidase activity and neurite outgrowth are abolished [241].

4.3 THE AMINOPEPTIDASE: THE BEGINNING AND THE END OF PROTEIN FUNCTION

A large number of peptides is released into the cytosol by intracellular proteolysis (Chapter 3). A small subset of them, which survives proteolytic attack, may exert biological functions. We will focus the discussion on the survival of the MHC-class I epitopes, which have been the subject of most experimental studies. These MHC class I-ligands, or their precursors, are created through a multistep pathway, starting with the breakdown of proteins by the proteasome. Kisselev et al. [282] discovered that the peptide products of mammalian proteasomes range in length from 3 to 22 amino acids, and their abundance decreases with increasing length, according to a log-normal distribution. About 70% of these peptides are too short (<7 amino acids) for antigen presentation. Only less than 15% of the released peptides are 8–9 amino acids residues in length [283]. There are cases where the proteasomes generate the final peptide ligand for MHC class I molecules [284, 285], while others carry N-terminal extensions [286, 287, 288, 289]. Concerning the latter, the aminopeptidase may correct the epitope's length and sequence by trimming the N-terminal extension of the precursor.

However, when the peptide reaches the cytosol, up to recently, it was assumed that it is readily converted into free amino acids by a number of ubiquitously distributed aminopeptidases [287, 290]. These are especially selective for oligopeptides, not only to remove the amino acid at the N-terminus [291] but also to remove dipeptides or tripeptides (dipeptidyl- and tripeptidylpeptidases) from polypeptide substrates [292; see 293]. Thus, intracellular peptides were not predicted to accumulate to detectable levels within the cell. However, hundreds of peptides derived from cytosolic proteins have been detected by mass spectrometric analysis of mouse tissues and human cell lines, raising the possibility that some peptides are much more stable than others within the cytosol of the cell [203, 294].

We are left with some questions, still awaiting convincing explanations: (i) how do bioactive peptides or their precursors (including the MHC class I epitopes) survive the attack by a variety of aminopeptidases (mainly metallo-, but also cysteine- and serine-peptidases [213], which are potentially able to convert any oligopeptide into free amino acids? (ii) Are bioactive fragments especially resistant, or have they been selected to escape degradation? If this is true, what is the molecular basis allowing the exclusion of this small set of bioactive products from complete hydrolysis?

Part of the answer to these intriguing questions was given by Gelman et al. [204]. Peptidomic analysis of the cytosol of human cell lines detected 203 peptides displaying free amino groups, out of which 69 were acetylated, therefore resistant to aminopeptidases. Interestingly, a recent analysis of the mouse brain peptidome found that approximately 50% of the cytosolic peptides represent the N- or C-termini of precursor proteins [294], thus suggesting that they are selectively produced and/or selectively retained.

It has been suggested that some peptides can escape complete degradation through transport into the endoplasmic reticulum via TAP [267, 295] or via peptide chaperones [259, 260]. Once the peptides arrive at the endoplasmic reticulum (ER), only the aminopeptidase ERAP1 [296] seems to be relevant, particularly for trimming epitope precursors that are extended at the amino terminus, thus preparing them to be of appropriate length to adjust to the groove of the MHC-class I molecule [289, 297]. Recently, it was shown that polymorphisms of the *ERAP1* gene, which encodes the endoplasmic reticulum peptidase, ERAP1, involved in peptide trimming before HLA class I presentation, are associated with a specific inflammatory arthritis (ankylosing spondylitis). This result provides strong evidence, connecting the pathology with the processing of antigenic peptides [298].

·　·　·　·　·

CHAPTER 5

Concluding Remarks

The protein components of eukaryotic cells face acute and chronic challenges to their integrity. Eukaryotic protein homeostasis enables healthy cell and organism development as well as aging, and protects against disease. Understanding the mechanisms that regulate protein synthesis and degradation is essential to respond to how the internal and environmental changes that happen, allow maintenance of homeostasis. Although during many decades the understanding of protein synthesis, as opposed to protein degradation, received most of the scientific efforts, recently new and revolutionary concepts have been conceived to explain how protein degradation is regulated. This is the case, for instance, for the N-end rule concept.

The N-termini of nascent proteins define their post-translational modifications, their half-lives, and consequently their functionality (N-end rule). The discovery of this rule, 25 years ago [299], was supposed to be true for a limited set of proteins. Recently, however, the N-end rule emerged as a major cellular process of signaling for degradation, which all proteins acquire when they are born. The degradation signals (degrons), targeted by the N-end rule pathways, are essential homeostasis elements of living organisms. An aminopeptidase is at the beginning of this process, since it has been shown that the full biological activity of a newly synthesized protein is preceded by the co-translational removal of Met by Met-amino-peptidases. After the removal of Met, about 80% of the nascent proteins are modified at their N-terminus (acetylation), not affecting their functionality, but signaling degradation by UPS [see 300, 301].

Therefore, the aminopeptidase is also present at the end of the full biological activity of a protein when the last peptide bond is hydrolyzed (Figure 12).

Life of multi-cellular organisms does not begin when proteins are expressed, folded and entwined, to give rise to the body's organization and function. An incessant process of communication and exchange of materials is essential, between the most remote cell of the body and the whole body, as well as between each organism and its environment, including a multitude of living species. These tasks are fulfilled not only by proteins but also by their proteolytic products. As described herein, limited proteolysis, despite its huge complexity far from being fully understood, is not a random process, but rather a well-organized enzymatic mechanism, deeply involved with homeostasis, which allows life to exist.

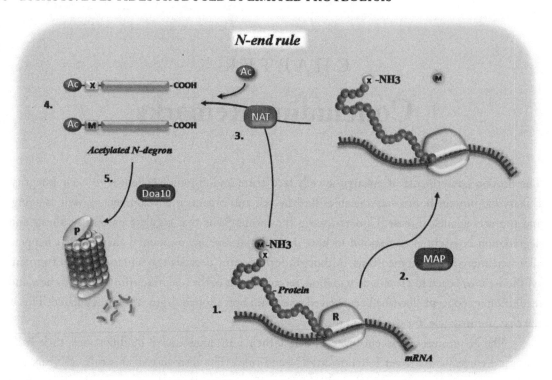

FIGURE 12: The N-end rule. Upon translation of mRNA into protein by the ribosome (R) (1), nascent polypeptide chains are cotranslationally processed by ribosome-associated methionine aminopeptidases (MAP) (2) and *N*-acetyltranferases (NAT) (3), generating conditional Ac-N-degrons (4) that can be targeted by the E3 ligase Doa10 to degradation by the proteasome (P) (5). The accessibility of acetylated amino termini (Ac) to Doa10 might link the stability of proteins to their folding or assembly state.

References

[1] Kyte J (1995). Structure in Protein Chemistry. Garland Publishing, Inc., New York & London, pp. 1.

[2] Langerak P, Russell P (2011). Regulatory networks integrating cell cycle control with DNA damage checkpoints and double-strand break repair. *Phil Trans R Soc B* 366: pp. 3562–71.

[3] Hartl FU, Bracher A, Hayer-Hartl M (2011). Molecular chaperones in protein folding and proteostasis. *Nature* 475: pp. 324–32.

[4] Herrmann J, Lerman LO, Lerman A (2007). Ubiquitin and ubiquitin-like proteins in protein regulation. *Circ Res* 100: pp. 1276–91.

[5] Korolchuk VI, Rubinsztein DC (2011). Regulation of autophagy by lysosomal positioning. *Autophagy* 7: pp. 927–8.

[6] Ratner S, Rittenberg D, Keston AS, Schoenheimer R (1940). Studies in protein metabolism XIV: the chemical interaction of dietary glycine and body proteins in rats. *J Biol Chem* 134: pp. 665–76.

[7] Guggenheim KY (1991). Rudolf Schoenheimer and the concept of the dynamic state of body constituents. *J Nutr* 121: pp. 1701–4.

[8] De Duve C, Gianetto R, Appelmans F, Wattiaux R (1953). Enzymic fraction of the mitochondria fraction. *Nature* 172: pp. 1143–4.

[9] Linderstøm-Lang (1950). Structure and enzymatic break-down of proteins. *Cold Spring Harb Symp Quant Biol* 14: pp. 117–26.

[10] Kirstein-Miles J, Morimoto RI (2010). Peptides signal mitochondrial stress. *Cell Metabol* 11: pp. 177–8.

[11] Fruton JS (1938). Protein structure and proteolytic enzymes. *Cold Spring Harb Symp Quant Biol* 6: pp. 50–7.

[12] Neurath H (1999). Proteolytic enzymes, past and future. *Proc Natl Acad Sci USA* 96: pp. 10962–3.

[13] Northrop JH, Kunitz M, Herriott RM (1938). Crystalline Enzymes. Columbia Univ. Press, New York.

[14] Rocha e Silva M, Beraldo WT, Rosenfeld G (1949). Bradykinin, a hypotensive and smooth

muscle stimulating factor released from plasma globulin by snake venoms and by trypsin. *Am J Physiol* 156: pp. 261–73.

[15]   Adibi SA, Mercer DW (1973). Protein digestion in human intestine as reflected in luminal, mucosal, and plasma amino acid concentrations after meals. *J Clin Invest* 52: pp. 1586–94.

[16]   Kau AL, Ahern PP, Griffin NW, Goodman AL, Gordon JI (2011). Human nutrition, the gut microbiome and the immune system. *Nature* 474: pp. 327–36.

[17]   Muegge BD, Kuczynski J, Knights D, Clemente JC, González A, Fontana L, Henrissat B, Knight R, Gordoni JI (2011). Diet drives convergence in gut microbiome functions across mammalian phylogeny and within humans. *Science* 332: pp. 970–4.

[18]   Faith JJ, McNulty NP, Rey FE, Gordon JI (2011). Predicting a human gut microbiota's response to diet in gnotobiotic mice. *Science* 333: pp. 101–4.

[19]   Kukkonen K, Kuitunen M, Haahtela T, Korpela R, Poussa T, Savilahti E (2010). High intestinal IgA associates with reduced risk of IgE-associated allergic diseases. *Pediatr Allergy Immunol* 21: pp. 67–73.

[20]   Faria AMC, Weiner HL (2005). Oral tolerance. *Immunol Rev* 206: pp. 232–59.

[21]   Vickery BP, Scurlock AM, Jones SM, Burks AW (2011). Mechanisms of immune tolerance relevant to food allergy. *J Allergy Clin Immunol* 127: pp. 576–84.

[22]   McCracken VJ, Lorenz RG (2001). The gastrointestinal ecosystem: a precarious alliance among epithelium, immunity and microbiota. *Cell Microbiol* 3: pp. 1–11.

[23]   Sharma R, Young C, Neu J (2010). Molecular modulation of intestinal epithelial barrier: contribution of microbiota. *J Biomed Biotech* 2010:ID 305879.

[24]   Shan L, Molberg Ø, Parrot I, Hausch F, Filiz F, Gray GM, Sollid LM, Khosla C (2002). Structural basis for gluten intolerance in celiac sprue. *Science* 297: pp. 2275–9.

[25]   McLachlan A, Cullis PG, Cornell HJ (2002). The use of extended amino acid motifs for focusing on toxic peptides in coeliac disease. *J Biochem Mol Biol Biophys* 6: pp. 319–24.

[26]   Hausch F, Shan L, Santiago NA, Gray GM, Khosla C (2002). Intestinal digestive resistance of immunodominant gliadin peptides. *Am J Physiol Gastrointest Liver Physiol* 283: pp. G996–1003.

[27]   Korhonen H, Pihlanto A (2006). Bioactive peptides: Production and functionality. *Intl Dairy J* 16: pp. 945–60.

[28]   Phelan M, Kerins D (2011). The potential role of milk-derived peptides in caridiovascular disease. *Food Funct* 2: pp. 153–67.

[29]   Jauhiainen T, Korpela T (2007). Milk peptides and blood pressure. *J Nutr* 137: pp. 825S–9S.

[30]   Nakamura Y, Yamamoto N, Sakai K, Takano T (1995). Antihypertensive effect of sour milk and peptides isolated from it that are inhibitors to angiotensin I-converting enzyme. *J Dairy Sci* 78: pp. 1253–7.

[31]   Ondetti MA, Cushman DW (1984). Angiotensin-converting enzyme inhibitors: biochemical properties and biological actions. *CRC Crit Rev Biochem* 16: pp. 381–411.

[32]   EFSA Scientific Report (2009). Review of the potential health impact of β-casomorphins and related peptides. *EFSA Scientific Report* 231: pp. 1–107.

[33]   The Washington Center for Clinical Research http://clinicaltrials.gov/ct2/show/NCT 00360919

[34]   Zoghbi S, Trompette A, Claustre J, El Homsi M, Garzón J, Jourdan G, Scoazec JY, Plaisancié P (2006). β-casomorphin-7 regulates the secretion and expression of gastrointestinal mucus through a μ-opioid pathway. *Am J Physiol Gastrointest Liver Physiol* 290: pp. G1105–13.

[35]   Hsieh C-C, Hernandez-Ledesma B, Jeong HJ, Park JH, de Lumen BO (2010). Complementary roles in cancer prevention: protease inhibitor makes the cancer preventive peptide lunasin bioavailable. *PLoS ONE* 5(1): p. e8890.

[36]   Hernández-Ledesma B, Hsieh CC, de Lumen BO (2011). Relationship between lunasin's sequence and its inhibitory activity of histones H3 and H4 acetylation. *Mol Nutr Food Res* 55: pp. 989–98.

[37]   Park JH, Jeong HJ, Lumen BO (2007). In vitro digestibility of the cancer-preventive soy peptides lunasin and BBI. *J Agric Food Chem* 55: pp. 10703–6.

[38]   Galvez AF, Chen N, Macasieb J, Lumen BO (2001). Chemopreventive property of a soybean Peptide (Lunasin). that binds to deacetylated histones and inhibits acetylation. *Cancer Res* 61: pp. 7473–8.

[39]   Coombes JL, Powrie F (2008). Dendritic cells in intestinal immune regulation. *Nat Rev Immunol* 8: pp. 435–46.

[40]   Dahan S, Roth-Walter F, Arnaboldi P, Agarwal S, Mayer L (2007). Epithelia: lymphocyte interactions in the gut. *Immunol Rev* 215: pp. 243–53.

[41]   Chehade M, Mayer L (2005). Oral tolerance and its relation to food hypersensitivities. *J Allergy Clin Immunol* 115: pp. 3–12.

[42]   Hershberg RM, Cho DH, Youakim A, Bradley MB, Lee JS, Framson PE, Nepom GT (1998). Highly polarized HLA class II antigen processing and presentation by human intestinal epithelial cells. *J Clin Invest* 102: pp. 792–803.

[43]   Iliev ID, Matteoli G, Rescigeno M (2007). The yin and yang of intestinal epithelial cells in controlling dendritic cell function. *J Exp Med* 204: pp. 2253–7.

[44]   Madani F, Lindberg S, Langel U, Futaki S, Gräslund A (2011). Mechanisms of cellular uptake of cell-penetrating peptides. *J Biophys* 2011:414729 doi:10.1155/2011/414729.

[45]   Sollid LM (2002). Coeliac disease: dissecting a complex inflammatory disorder. *Nat Rev Immunol* 2: pp. 647–55.

[46]   Lundin KE, Gjertsen HA, Scott LM, Sollid LM, Thorsby E (1994). Function of DQ2 and DQ8 as HLA susceptibility molecules in celiac disease. *Hum Immunol* 41: pp. 24–7.

[47]   Trier JS (1991). Celiac sprue. *N Engl J Med* 325: pp. 1709–19.

[48]   Shan L, Qiao S, Arentz-Hansen H, Molberg Ø, Gray GM, Sollid LM, Khosla C (2005). Peptides from gluten: implications for celiac sprue. *J Proteome Res* 4: pp. 1732–41.

[49]   Kim CY, Quarsten H, Bergseng E, Khosla C, Sollid LM (2004). Structural basis for HLA-DQ2-mediated presentation of gluten epitopes in celiac disease. *Proc Natl Acad Sci USA* 101: pp. 4175–9.

[50]   Stepniak D, Wiesner M, de Ru AH, Moustakas AK, Drijfhout JW, Papadopoulos GK, van Veelen P, Koning F (2008). Large-scale characterization of natural ligands explains the unique gluten-binding properties of HLA-DQ2. *J Immunol* 180: pp. 3268–78.

[51]   Molberg Ø, Mcadam SN, Körner R, Quarsten H, Kristiansen C, Madsen L, Fugger L, Scott H, Norén O, Roepstorff P, Lundin KE, Sjöström H, Sollid LM (1998). Tissue trans-glutaminase selectivity modifies gliadin peptides that are recognized by gut-derived T cells in celiac disease. *Nat Med* 4: pp. 713–7.

[52]   Jin X, Stamnaes J, Klöck C, DiRaimondo TR, Sollid LM, Khosla C (2011). Activation of extracellular transglutaminase 2 by thioredoxin. *J Biol Chem* 286: pp. 37866–73.

[53]   Marti T, Molberg Ø, Li Q, Gray GM, Khosla C, Sollid LM (2005). Prolyl endopeptidase-mediated destruction of T cell epitopes in whole gluten: Chemical and immunological characterization. *J Pharmcol Extl Therap* 312: pp. 19–26.

[54]   Fuhrmann G, Leroux J-C (2011). In vivo fluorescence imaging of exogenous enzyme activity in the gastrointestinal tract. *Proc Natl Acad Sci USA* 108: pp. 9032–7.

[55]   Camargo AC, Caldo H, Reis ML (1979). Susceptibility of a peptide brain endo-oligo-peptidases derived from bradykinin to hydrolysis and pancreatic proteinases. *J Biol Chem* 254: pp. 5304–7.

[56]   Cicilini MA, Caldo H, Berti JD, Camargo AC (1977). Rabbit tissue peptidases that hydro-lyse the peptide hormone bradykinin. Differences in enzymic properties and concentration in rabbit tissues. *Biochem J* 163: pp. 433–9.

[57]   Rea D, Fulöp V (2011). Prolyl oligopeptidase structure and dynamics. *CNS Neurol Disord Drug Targets* 10: pp. 306–10.

[58]   Tigerstedt R, Bergman PG (1898). Kidney and circulation. *Scand Arch Physiol* 8: pp. 223–71.

[59]   Goldblatt H, Lynch J, Ramon F, Summerville WW (1934). Studies on experimental hyper-tension I. The production of persistent elevation of systolic blood pressure by means of renal ischemia. *J Exp Med* 59: pp. 347–79.

[60]   Braun-Menéndez E, Fasciolo JC, Leloir LF, Muñoz JM (1939). La sustancia hipertensora de la sangre del riñón isquemiado. *Rev Soc Arg Biol* 15: pp. 420–5.

[61] Page IH, Helmer OM (1940). A crystalline pressor substance (angiotonin). resulting from the reaction between rennin and rennin-activator. *J Exp Med* 71: pp. 29–42.

[62] Schmaier AH (2002). The plasma kallikrein-kinin system counterbalances the renin–angiotensin system. *J Clin Invest* 109: pp. 1007–9.

[63] Elliot DF, Horton EW, Lewis GP (1960). Actions of pure bradykinin. *J Physiol* 153: pp. 473–80.

[64] Schechter AN, Dean A, Goldberger RF, Eds (1984). The Impact of Protein Chemistry on the Biomedical Sciences. Academic Press, Inc., NY.

[65] Nguyen G, Delarue F, Burcklé C, Bouzhir L, Giller T, Sraer JD (2002). Pivotal role of the renin/prorenin receptor in angiotensin II production and cellular responses to renin. *J Clin Invest* 109: pp. 1417–27.

[66] Nguyen G, Muller DN (2010). The biology of the prorenin receptor. *J Am Soc Nephrol* 21: pp. 18–23.

[67] Lalmanach G, Naudin C, Lecaille F, Fritz H (2010). Kininogens: More than cysteine protease inhibitors and kinin precursors. *Biochimie* 92: pp. 1568–79.

[68] Sainz IM, Pixley RA, Colman RW (2007). Fifty years of research on the plasma kallikrein-kinin system:from protein structure and function to cell biology and in-vivo pathophysiology. *Thromb Haemost* 98: pp. 77–83.

[69] Sealey JE, White RP, Laragh JH, Rubin AL (1977). Plasma prorenin and renin in anephric patients. *Circ Res* 41: pp. 17–21.

[70] Andreasen PA, Egelund R, Petersen HH (2000). The plasminogen activation system in tumor growth, invasion, and metastasis. *Cell Mol Life Sci* 57: pp. 25–40.

[71] Hunyady L, Catt KJ (2006). Pleiotropic AT1 receptor signaling pathways mediating physiological and pathogenic actions of angiotensin II. *Mol Endocrinol* 20: pp. 953–70.

[72] Zhu L, Carretero OA, Liao TD, Harding P, Li H, Sumners C, Yang XP (2010). Role of prolylcarboxypeptidase in angiotensin II type 2 receptor-mediated bradykinin release in mouse coronary artery endothelial cells. *Hypertension* 56: pp. 384–90.

[73] Bernstein KE (2002). Two ACEs and a heart. *Nature* 417: pp. 799–802.

[74] Ferrario CM, Varagig J (2010). The Ang-(1-7)/ACE2/mas axis in the regulation of nephron function. *Am J Physiol Renal Physiol* 298: pp. F1297–305.

[75] Schmaier AH (2003). The kallikrein-kinin and the renin-angiotensin systems have a multilayered interaction. *Am J Physiol Regul Integr Comp Physiol* 285: pp. R1–13.

[76] Schmaier AH (2008). Assembly, activation, and physiologic influence of the plasma kallikrein/kinin system. *Int Immunopharmacol* 8: pp. 161–5.

[77] Kaplan AP, Joseph K, Silverberg M (2002). Pathways for bradykinin formation and inflammatory disease. *J Allergy Clin Immunol* 109: pp. 195–209.

[78] Moreau ME, Garbacki N, Molinaro G, Brown NJ, Marceau F, Adam A (2005). The

kallikrein-kinin system: current and future pharmacological targets. *J Pharmacol Sci* 99: pp. 6–38.

[79]   Kaplan AP, Silverberg M (1987). The coagulation-kinin pathway of human plasma. *Blood* 70: pp. 1–15.

[80]   Marceau F, Regoli D (2004). Bradykinin receptor ligands: therapeutic perspectives. *Nat Rev Drug Discov* 3: pp. 845–52.

[81]   Miller DK, Ayala JM, Egger LA, Raju SM, Yamin TT, Ding GJ, Gaffney EP, Howard AD, Palyha OC, Rolando AM, Salley JP, Thornberry NA, Weidner JR, Williams JH, Chapman KT, Jackson J, Kostura MJ, Limjucoll G, Molineaux SM, Mumford RA, Calaycay JR (1993). Purification and characterization of active human interleukin-1 beta-converting enzyme from THP.1 monocytic cells. *J Biol Chem* 268:v18062–9.

[82]   Stutz A, Golenbock DT, Latz E (2009). Science in medicine Inflammasomes: too big to miss. *J Clin Invest* 119: pp. 3502–11.

[83]   Pappu R, Ramirez-Carrozzi V, Arivazhagan S (2011). The interleukin-17 cytokine family: critical players in host defence and inflammatory diseases. *Immunology* 134: pp. 8–16.

[84]   Heutinck KM, ten Berge IJ, Hack CE, Hamann J, Rowshani AT (2010). Serine proteases of the human immune system in health and disease. *Mol Immunol* 47: pp. 1943–55.

[85]   Marceau F, Hess JF, Bacharov DR (1998). The B1 receptors for kinins. *Pharmacol Rev* 50: pp. 357–86.

[86]   Kuhr F, Lowry J, Zhang Y, Brovkovych V, Skidgel RA (2010). Differential regulation of inducible and endothelial nitric oxide synthase by kinin B1 and B2 receptors. *Neuropeptides* 44: pp. 145–54.

[87]   Yang HY, Erdös EG, Levin Y (1970). A dipeptidyl carboxypeptidase that converts angiotensin I and inactivates bradykinin. *Biochem Biophys Acta* 214: pp. 374–6.

[88]   Ng KK, Vane JR (1970). Some properties of angiotensin converting enzyme in the lung in vivo. *Nature* 225: pp. 1142–4.

[89]   Camargo AC, Ianzer DA, Guerreiro JR, Serrano SM (2012). Bradykinin-potentiating peptides: beyond captopril. *Toxicon* 59: pp. 516–23.

[90]   Cotton J, Hayashi MA, Curiasse P, Vazeux G, Ianzer D, Camargo AC, Dive V (2002). Selective inhibition of the C-domain of angiotensin I converting enzyme by bradykinin potentiating peptides. *Biochemistry* 41: pp. 6065–71.

[91]   Gavras H, Brunner HR, Laragh JH, Sealey JE, Gavras I, Vukovich RA (1974). An angiotensin converting-enzyme inhibitor to identify and treat vasoconstrictor and volume factors in hypertensive patients. *N Engl J Med* 291: pp. 817–21.

[92]   Ondetti MA, Rubin B, Cushman DW (1977). Design of specific inhibitors of angiotensin-converting enzyme: new class of orally active antihypertensive agents. *Science* 196: pp. 441–4.

[93]   Donoghue M, Hsieh F, Baronas E, Godbout K, Gosselin M, Stagliano N, Donovan M, Woolf B, Robison K, Jeyaseelan R, Breitbart RE, Acton S (2000). A novel engiotensin-converting enzyme-related caboxypeptidase (ACE2). converts angiotensin I to angiotensin 1–9. *Circ Res* 87: pp. e1–9.

[94]   Siragy HM, Jaffa AA, Margolius HS, Carey RM (1996). Renin-angiotensin sustem modulates renal bradykinin production. *Am J Physiol Integr Comp Physiol* 271: pp. R1090–5.

[95]   Bader M, Ganten D (2008). Update on tissue rennin-angiotensin systems. *J Mol Med* 86: pp. 615–21.

[96]   Konno K, Palma MS, Hirata IY, Juliano MA, Juliano L, Yasuhara T (2002). Identification of bradykinins in solitary wasp venoms. *Toxicon* 40: pp. 309–12.

[97]   Cornell MJ, Williams TA, Lamango NS, Coates D, Corvol P, Soubrier F, Hoheisel J, Lehrach H (1995). Cloning and expression of an evolutionary conserved single-domain angiotensin converting enzyme from *Drosophila melanogaster*. *J Biol Chem* 270: pp. 13613–9.

[98]   Martinez-Torres A, Miledi R (2006). Expression of *Caenorhabditis elegans* neurotransmitter receptors and ion channels in *Xenopus* oocytes. *Proc Natl Acad Sci USA* 103: pp. 5120–4.

[99]   Schoenheimer R (1942). The Dynamic State of Body Constituents. Harvard University Press, Cambridge, MA.

[100]   Yen HC, Xu Q, Chou DM, Zhao Z, Elledge SJ (2008). Global protein stability profiling in mammalian cells. *Science* 322: pp. 918–23.

[101]   Etlinger JD, Goldberg AL (1977). A soluble ATP-dependent proteolytic system responsible for the degradation of abnormal proteins in reticulocytes. *Proc Natl Acad Sci USA* 74: pp. 54–8.

[102]   Hershko A, et al., (1978). In Protein Turnover and Lysosome Function (Segal HL and Doyle DJ, Eds), Academic Press, New York, pp. 149–69.

[103]   Ciechanover A (2005). Intracellular protein degradation: from a vague idea, through the lysosome and the ubiquitin–proteasome system, and onto human diseases and drug targeting. *Angew Chem Int Ed* 44: pp. 5944–67.

[104]   Ciechanover A (2005). Proteolysis: from the lysosome to ubiquitin and the proteasome. *Nat Rev Mol Cell Biol* 6: pp. 79–87.

[105]   Sorokin AV, Kim ER, Ovchinnikov LP (2009). Proteasome system of protein degradation and processing. *Biochemistry (Mosc)*. 74: pp. 1411–42.

[106]   Kisselev AF, Songyang Z, Goldberg AL (2000). Why does threonine, and not serine, function as the active site nucleophile in proteasomes? *J Biol Chem* 275: pp. 14831–7.

[107]   Heinemeyer W, Fischer M, Krimmer T, Stachon U, Wolf DH (1997). The active sites of the eukaryotic 20S proteasome and their involvement in subunit precursor processing. *J Biol Chem* 272: pp. 25200–9.

[108]   Arendt CS, Hochstrasser (1997). Identification of the yeast 20S proteasome catalytic

centers and subunit interactions required for active-site formation. *Proc Natl Acad Sci USA* 94: pp. 7156–61.

[109]   Kloetzel PM (2001). Antigen processing by the proteasome. *Nat Rev Mol Cell Biol* 2: pp. 179–87.

[110]   Schmidtke G, Eggers M, Ruppert T, Groettrup M, Koszinowski UH, Kloetzel PM (1998). Inactivation of a defined active site in the mouse 20S proteasome complex enhances major histocompatibility complex class I antigen presentation of a murine cytomegalovirus protein. *J Exp Med* 187: pp. 1641–6.

[111]   Nussbaum AK, Dick TP, Keilholz W, Schirle M, Stevanocic S, Dietz K, Heinemeyer W, Groll M, Wolf DH, Huber R, Rammensee HG, Schild H (1998). Cleavage motifs of the yeast 20S proteasome beta subunits deduced from digests of enolase 1. *Proc Natl Acad Sci USA* 95: pp. 12504–9.

[112]   Brooks P, Fuertes G, Murray RZ, Bose S, Knecht E, Rechsteiner MC, Hendil KB, Tanaka K, Dyson J, Rivett J (2000). Subcellular localization of proteasomes and their regulatory complexes in mammalian cells. *Biochem J* 346: pp. 155–61.

[113]   Braun BC, Glickman M, Kraft R, Dahlmann B, Kloetzel PM, Finley D, Schmidt M (1999). The base of the proteasome regulatory particle exhibits chaperone-like activity. *Nat Cell Biol* 1: pp. 221–6.

[114]   Strickland E, Hakala K, Thomas PJ, DeMartino GN (2000). Recognition of misfolding proteins by PA700, the regulatory subcomplex of the 26S proteasome. *J Biol Chem* 275: pp. 5565–72.

[115]   Groll M, Bajorek M, Köhler A, Moroder L, Rubin DM, Huber R, Glickman MH, Finley D (2000). A gated channel into the proteasome core particle. *Nat Struct Biol* 7: pp. 1062–7.

[116]   Köhler A, Bajorek M, Groll M, Moroder L, Rubin DM, Huber R, Glickman MH, Finley D (2001). The substrate translocation channel of the proteasome. *Biochimie* 83: pp. 325–32.

[117]   Hershko A, Leshinsky E, Ganoth D, Heller H (1984). ATP-dependent degradation of ubiquitin-protein conjugates. *Proc Natl Acad Sci USA* 81: pp. 1619–23.

[118]   Ravid T, Hochstrasser M (2008). Diversity of degradation signals in the ubiquitin-proteasome system. *Nat Rev Molec Cell Biol* 9: pp. 679–90.

[119]   Varshavsky A (2005). Regulated protein degradation. *Trends Biochem Sci* 30: pp. 283–6.

[120]   Yao T, Cohen RE (2002). A cryptic protease couples deubiquitination and degradation by the proteasome. *Nature* 419: pp. 403–7.

[121]   Babbitt BP, Allen PM, Matsueda G, Haber E, Unanue ER (1985). Binding of immunogenic peptides to Ia histocompatibility molecules. *Nature* 317: pp. 359–61.

[122]   Bodmer HC, Bastin JM, Askonas BA, Townsend AR (1989). Influenza-specific cytotoxic

T-cell recognition is inhibited by peptides unrelated in both sequence and MHC restriction. *Immnunology* 66: pp. 163–9.

[123]   Gopalakrishnan B, Roques BP (1992). Do antigenic peptides have a unique sense of direction inside the MHC binding groove? A molecular modelling study. *FEBS Lett* 303: pp. 224–8.

[124]   Starck SR, Shastri N (2011). Non-conventional sources of peptides presented by MHC class I. *Cell Mol Life Sci* 68: pp. 1471–9.

[125]   Larsen JE, Lund O, Nielsen M (2002). Improved method for predicting linear B-cell epitopes. *Immunome Research* 2: p. 2.

[126]   Liu T, Liu W, Song Z, Jiao C, Zhu M, Wang X (2009). Computational prediction of the specificities of proteasome interaction with antigen protein. *Cell Mol Immunol* 6: pp. 135–42.

[127]   Denzin LK, Cresswell P (1995). HLA-DM induces CLIP dissociation from MHC class II alpha beta dimers and facilitates peptide loading. *Cell* 82: pp. 155–85.

[128]   Leddon SA, Sant AJ (2010). Generation of MHC class II: peptide ligands for CD4 T cell allorecognition of MHC Class II molecules. *Curr Opin Organ Trans* 15: pp. 505–11.

[129]   Sadegh-Nasseri S, Natarajan S, Chou CL, Hartman IZ, Narayan K, Kim A (2010). Conformational heterogeneity of MHC class II induced upon binding to different peptides is a key regulator in antigen presentation and epitope selection. *Immunol Res* 47: pp. 56–64.

[130]   Reits E, Griekspoor A, Neijssen J, Groothuis T, Jalink K, van Veelen P, Janssen H, Calafat J, Drijfhout JW, Neefjes J (2003). Peptide diffusion, protection, and degradation in nuclear and cytoplasmic compartments before antigen presentation by MHC class I. *Immunity* 18: pp. 97–108.

[131]   Blanchard N, Shastri N (2008). Coping with loss of perfection in the MHC class I peptide repertoire. *Curr Opin Immunol* 20: pp. 82–8.

[132]   http://www.ddg-pharmfac.net/antijen/AntiJen/antijenhomepage.htm

[133]   Groettrup M, Soza A, Kuckelkorn U, Kloetzel PM (1996). Peptide antigen production by the proteasome: complexity provides efficiency. *Immunol Today* 17: pp. 429–35.

[134]   Aki M, Shimbara N, Takashina M, Akiyama K, Kagawa S, Tamura T, Tanahashi N, Yoshimura T, Tanaka K, Ichihara A (1994). Interferon-gamma induces different subunit organizations and functional diversity of proteasomes. *J Biochem* 115: pp. 257–69.

[135]   Sijts EJA, Kloetzel MPM (2011). The role of the proteasome in the generation of MHC class I ligands and immune responses. *Cell Mol Life Sci* 68: pp. 1491–502.

[136]   Falk K, Rötschke O, Deres K, Metzger J, Jung G, Rammensee HG (1991). Identification of naturally processed viral nonapeptides allows their quantification in infected cells and suggests an allele-specific T cell epitope forecast. *J Exp Med* 174: pp. 425–34.

[137] Rammensee HG, Friede T, Stevanoviíc S (1995). MHC ligands and peptide motifs: first listing. *Immunogenetics* 41: pp. 178–228.

[138] Reits EA, Vos JC, Grommé M, Neefjes J (2000). The major substrates for TAP *in vivo* are derived from newly synthetized proteins. *Nature* 404: pp. 774–8.

[139] Momburg F, Hämmerling GJ (1998). Generation and TAP-mediated transport of peptides for major histocompatibility complex class I molecules. *Adv Immunol* 68: pp. 191–256.

[140] Pietra G, Romagnani C, Manzini C, Moretta L, Mingari MC (2002). Conformational heterogeneity of MHC class II induced upon binding to different peptides is a key regulator in antigen presentation and epitope selection. *J Biomed Biotechnol* 2010:ID 907092.

[141] Evavold BD, Allen PM (1991). Separation of IL-4 production from Th cell proliferation by an altered T cell receptor ligand. *Science* 252: pp. 1308–10.

[142] Unanue ER (2011). Altered peptide ligands make their entrance. *J Immunol* 186: pp. 7–8.

[143] Modlin IM, Champaneria MC, Bornschein J, Kidd M (2006). Evolution of the diffuse neuroendocrine system—clear cells and cloudy origins. *Neuroendocrinol* 84: pp. 69–82.

[144] Day R, Salzet M (2002). The neuroendocrine phenotype, cellular plasticity, and the search for genetic switches: redefining the diffuse neuroendocrine system. *Neuro Endocrinol Lett* 23: pp. 447–51.

[145] Reichlin S (1993). Neuroendocrine-immune interactions. *N Engl J Med* 329: pp. 1246–53.

[146] Steiner D (2011). Adventures with insulin in the islets of Langerhans. *J Biol Chem* 286: pp. 17399–421.

[147] de Bree FM (2000). Trafficking of the vasopressin and oxytocin prohormone through the regulated secretory pathway. *J Neuroendocrinol* 12: pp. 589–94.

[148] Banting FG, Best CH (1922). Pancreatic extracts. *J Lab Clin Med* 7: pp. 464–72.

[149] Sanger F, Tuppy H (1951). The amino-acid sequence in the phenylalanyl chain of insulin. I. The identification of lower peptides from partial hydrolysates. *Biochem* 49: pp. 463–81.

[150] Humbel RE (1965). Biosynthesis of the two chains of insulin. *Proc Natl Acad Sci USA* 53: pp. 835–59.

[151] Steiner DF, Cunningham D, Spigelman L, Aten B (1967). Insulin biosynthesis: evidence for a precursor. *Science* 157: pp. 697–700.

[152] Zühlke H, Steiner DF, Lernmark A, Lipsey C (1976). Carboxypeptidase B-like and trypsin-like activities in isolated rat pancreatic islets. *Ciba Found Symp* 41: pp. 183–95.

[153] Naggert JK, Fricker LD, Varlamov O, Nishina PM, Rouille Y, Steiner DF, Carroll RJ, Paigen BJ, Leiter EH (1995). Hyperproinsulinaemia in obese fat/fat mice associated with a carboxypeptidase E mutation which reduces enzyme activity. *Nat Genet* 10: pp. 135–42.

[154] Davidson HW, Hutton JC (1987). The insulin-secretory-granule carboxypeptidase H: Purification and demonstration of involvement in proinsulin processing. *Biochem J* 245: pp. 575–82.

[155] Chance RE, Ellis RM, Bromer WW (1968). Porcine proinsulin: characterization and amino acid sequence. *Science* 161: pp. 165–7.

[156] Hubbard SJ, Beynon RJ, Thornton JM (1998). Assessment of conformational parameters as predictors of limited proteolytic sites in native protein structures. *Protein Engineering* 11: pp. 349–59.

[157] Kazanov MD, Igarashi Y, Eroshkin AM, Cieplak P, Ratnikov B, Zhang Y, Li Z, Godzik A, Osterman AL, Smith JW (2011). Structural determinants of limited proteolysis. *J Proteome Res* 10: pp. 3642–51.

[158] Jamieson JD, Palade GE (1967). Intracellular transport of secretory proteins in the pancreatic exocrine cell. I. Role of the peripheral elements of the Golgi complex. *J Cell Biol* 34: pp. 577–615.

[159] Blobel G, Dobberstein B (1975). Transfer of proteins across membranes. I. Presence of proteolytically processed and unprocessed nascent immunoglobulin light chains on membrane-bound ribosomes of murine myeloma. *J Cell Biol* 67: pp. 835–51.

[160] Steiner DF, Clark JL, Nolan C, Rubenstein AH, Margoliash E, Aten B, Oyer PE (1969). Proinsulin and the biosynthesis of insulin. *Recent Prog Horm Res* 25: pp. 207–82.

[161] Orci L, Ravazzola M, Amherdt M, Madsen O, Vassalli JD, Perrelet A (1985). Direct identification of prohormone conversion site in insulin-secreting cells. *Cell* 42: pp. 671–81.

[162] Davidson HW, Peshavaria M, Hutton JC (1987). Proteolytic conversion of proinsulin into insulin. Identification of a Ca^{2+}-dependent acidic endopeptidase in isolated insulin-secretory granules. *Biochem J* 246: pp. 279–86.

[163] Fricker LD, Snyder SH (1982). Enkephalin convertase: purification and characterization of a specific enkephalin-synthesizing carboxypeptidase localized to adrenal chromaffin granules. *Proc Natl Acad Sci USA* 79: pp. 3886–90.

[164] Davidson HW, Rhodes CJ, Hutton JC (1988). Intraorganellar calcium and pH control proinsulin cleavage in the pancreatic beta cell via two distinct site-specific endopeptidases. *Nature* 333: pp. 93–6.

[165] Bennett DL, Bailyes EM, Nielsen E, Guest PC, Rutherford NG, Arden SD, Hutton JC (1992). Identification of the typ2e proinsulin processing endopeptidase as PC2, a member of the eukaryote subtilisin family. *J Biol Chem* 267: pp. 15229–36.

[166] Julius D, Brake A, Blair L, Kunisawa R, Thorner J (1984). Isolation of the putative structural gene for the lysine-arginine-cleaving endopeptidase required for processing of yeast prepro-alpha-factor. *Cell* 37: pp. 1075–89.

[167] Smeekens SP, Avruch AS, LaMendola J, Chan SJ, Steiner DF (1991). Identification of a cDNA encoding a second putative prohormone convertase related to PC2 in AtT20 cells and islets of Langerhans. *Proc Natl Acad Sci USA* 88: pp. 340–4.

[168] Chretien M, Gasper L, Benjannet S, Mbikay M, Lazure C, Seidah NG (1991). From

POMC to functional diversity of neural peptides: the key importance of the convertases. *Trans Am Clin Climatol Assoc* 102: pp. 195–224.

[169] Hook V, Funkelstein L, Lu D, Bark S, Wegrzyn J, Hwang SR (2008). Proteases for processing proneuropeptides into peptide neurotransmitters and hormones. *Annu Rev Pharmacol Toxicol* 48: pp. 393–423.

[170] Bailyes EM, Shennan KI, Seal AJ, Smeekens SP, Steiner DF, Hutton JC, Docherty K (1992). A member of the eukaryotic subtilisin family (PC3). has the enzymic properties of the type 1 proinsulin-converting endopetidase. *Biochem J* 285: pp. 391–4.

[171] Lu WD, Asmus K, Hwang SR, Li S, Woods VL, Jr., Hook V (2009). Differetial accessibilities of dibasic prohormone processing sites of proenkephalin to the aqueous environment revealed by H-D exchange mass spectrometry. *Biochemistry* 48: pp. 1604–12.

[172] Mains RE, Bloomquist BT, Eiper BA (1991). Manipulation of neuropeptide biosynthesis through the expression of antisense RNA for peptidylglycine alpha-amidating monooxygenase. *Mol Endocrinol* 5: pp. 187–93.

[173] Seidah NG (2011). What lies ahead for the proprotein convertases? *Ann NY Acad Sci* 1220: pp. 149–61.

[174] Seidah NG, Mayer G, Zaid A, Rousselet E, Nassoury N, Poirier S, Essalmani R, Prat A (2008). The activation and physiological functions of the proprotein convertases. *Int J Biochem Cell Biol* 40: pp. 1111–25.

[175] Camargo AC, Ribeiro MJ, Schwartz WN (1985). Conversion and inactivation of opioid-peptides by rabbit brain endo-oligopeptidase A. *Biochem Biophys Res Commun* 130: pp. 932–8.

[176] Camargo AC, Oliveira EB, Toffoletto O, Metters KM, Rossier J (1987). Brain endo-oligopeptidase A, a putative enkephalin converting enzyme. *J Neurochem* 48: pp. 1258–63.

[177] Ferro ES, Tambourgi DV, Gobersztejn F, Gomes MD, Sucupira M, Armelin MC, Kipnis TL, Camargo AC (1993). Secretion of a neuropeptide-metabolizing enzyme similar to endopeptidase 22.19 by glioma C6 cells. *Biochem Biophys Res Commun* 191: pp. 275–81.

[178] Le Merrer J, Becker JA, Befort K, Kieffer BL (2009). Reward processing by the opioid system in the brain. *Physiol Rev* 89: pp. 1379–412.

[179] Zhu X, Orci L, Carroll R, Norrbom C, Ravazzola M, Steiner DF (2002). Severe block in processing of proinsulin to insulin accompanied by elevation of des-64,65 proinsulin intermediates in islets of mice lacking prohormone convertase 1/3. *Proc Natl Acad Sci USA* 99: pp. 10299–304.

[180] Hoshino A, Lindberg I (2012). Peptide Biosynthesis: Prohormone Convertases 1/3 and 2. Colloquium Series on Neuropeptides (Fricker L and Devi L, Eds), Morgan Claypool Publisher.

[181] Prichard LE, Turnbull AV, White A (2002). Pro-opiomelanocortin processing in the hypo-thalamus: impact on melanocortin signalling and obesity. *J Endocrinol* 172: pp. 411–21.

[182] Hruby VJ, Sharma SD, Toth K, Jaw JY, al-Obeidi F, Sawyer TK, Hadley ME (1993). De-sign, synthesis, and conformation of superpotent and prolonged acting melanotropins. *Ann NY Acad Sci* 680: pp. 51–63.

[183] Funkelstein L, Beinfeld M, Minokadeh A, Zadina J, Hook V (2010). Unique biological function of cathepsin L in secretory vesicles for biosynthesis of neuropeptides. *Neuropept* 44: pp. 457–66.

[184] Zhang X, Che FY, Berezniuk I, Sonmez K, Toll L, Fricker LD (2008). Peptidomics of Cpe(fat/fat). mouse brain regions: implications for neuropeptide processing. *J Neurochem* 107: pp. 1596–613.

[185] Murayama N, Hayashi MA, Ohi H, Ferreira LA, Hermann VV, Saito H, Fujita Y, Higuchi S, Fernandes BL, Yamane T, Camargo AC (1997). Cloning and sequence analysis of a Both-rops jararaca cDNA encoding a precursor of seven bradykinin-potentiating peptides and a C-type natriuretic peptide. *Proc Natl Acad Sci USA* 94: pp. 1189–93.

[186] Hayashi MA, Murbach AF, Ianzer D, Portaro FC, Prezoto BC, Fernandes BL, Silveira PF, Silva CA, Pires RS, Britto LR, Dive V, Camargo AC (2003). The C-type natriuretic peptide precursor of snake brain contains highly specific inhibitors of the angiotensin-converting enzyme. *J Neurochem* 85: pp. 969–77.

[187] Dungan HM, Clifton DK, Steiner RA (2006). Kisspeptin neurons as central processors in the regulation of gonadotropin-releasing hormone secretion. *Endocrinology* 147: pp. 1154–8.

[188] Navarro VM, Gottsch ML, Chavkin C, Okamura H, Clifton DK, Steiner RA (2009). Regulation of gonatropin-releasing hormone secretion by kisspeptin/dynorphin/neurokinin B neurons in the arcuate nucleus of the mouse. *J Neurosci* 29 pp.:11859–66.

[189] Ducancel F, Vilborg M, Dupont C, Lajeunesse E, Wollberg Z, Bdohla A, Kochva E, Bou-lain JC, Ménez A (1993). Cloning and sequence analysis of cDNAs encoding precursors of sarafotoxins. *J Biol Chem* 268: pp. 3052–5.

[190] Elphick MR (2010). NG peptides: A novel family of neurolysin-associated neuropeptides. *Gene* 458: pp. 20–6.

[191] Elphick M, Rowe ML (2009). NGFFFamide and echinotocin: structurally unrelated myo-active neuropeptides derived from neurolysin-containing precursors in sea urchin. *J Exp Biol* 212: pp. 1067–77.

[192] Reinscheid RK (2007). Phylogenetic appearance of neuropeptide S precursor proteins in tetrapods. *Peptides* 28: pp. 830–7.

[193] De Baets G, Reumers J, Blanco JD, Dopazo J, Schymkowitz J, Rousseau F (2011). An

evolutionary trade-off between protein turnoverrate and protein aggregation favors a higher aggregation propensity in fast degrading proteins. *PLoS Comput Biol* 7: p. e1002090.

[194]   Young L, Leonhard K, Tatsuta T, Trowsdale J, Langer T (2001). Role of the ABC transporter Mdl1 in peptide export from mitochondria. *Science* 291: pp. 2135–8.

[195]   Haynes CM, Yang Y, Blais SP, Neubert TA, Ron D (2010). The matrix peptide exporter HAF-1 signals a mitochondrial UPR by activating the transcription factor ZC376.7 in *C. elegans*. *Mol Cell* 37: pp. 529–40.

[196]   Haynes CM, Petrova K, Benedetti C, Yang Y, Ron D (2007). ClpP mediates activation of a mitochondrial unfolded protein response in *C. elegans*. *Dev Cell* 13: pp. 467–80.

[197]   Galindo MI, Pueyo JI, Fouix S, Bishop AS, Couso JP (2007). Peptides encoded by short ORFs control development and define a new eukaryotic gene family. *Plos Biol* 5: p. e106.

[198]   Kondo T, Hashimoto Y, Kato K, Inagaki S, Hayashi S, Kageyama Y (2007). Small peptides regulators of actin-based cell morphogenesis encoded by a polycistronic mRNA. *Nat Cell Biol* 9: pp. 660–5.

[199]   Kondo T, Plaza S, Zanet J, Benrabah E, Valenti P, Hashimoto Y, Kobayashi S, Payre F, Kageyama Y (2010). Small peptides switch the transcriptional activity of Shavenbaby during *Drosophila* embryogenesis. *Science* 329: pp. 336–9.

[200]   Carninci P, et al., (2005). The transcriptional landscape of the mammalian genome. *Science* 309: pp. 1559–63.

[201]   Amaral PP, Mattick JS (2008). Noncoding RNA in development. *Mamm Genome* 19: pp. 454–92.

[202]   London N, Raveh B, Cohen E, Fathi G, Schueler-Furman O (2011). Rosetta FlexPepDock web server—high resolution modeling of peptide–protein interactions. *Nucl Acids Res* 39: pp. W249–53.

[203]   Fricker FD (2010). Analysis of mouse brain peptides using mass spectrometry-based peptidomics: implications for novel functions ranging from non-classical neuropeptides to microproteins. *Mol Biosyst* 6: pp. 1355–65.

[204]   Gelman JS, Fricker LD (2010). Hemopressin and other bioactive peptides from cytosolic proteins: are these non-classical neuropeptides? *The AAPS Journal* 12: pp. 279–89.

[205]   Ferro ES, Hyslop S, Camargo AC (2004). Intracellular peptides as putative natural regulators of protein interactions. *J Neurochem* 91: pp. 769–77.

[206]   London N, Movshovitz-Attias D, Schueler-Furman O (2010). The structural basis of peptide-protein binding strategies. *Structure* 18: pp. 188–99.

[207]   He B, Minges JT, Lee LW, Wilson EM (2002). The FXXLF motif mediates androgen receptor-specific interactions with coregulators. *J Biol Chem* 277: pp. 10226–35.

[208]   Akiva E, Friedlanderb G, Itzhakic Z, Margalit H (2012). A dynamic view of domain-motif interactions. *PLoS Comput Biol* 8: p. e1002341.

[209] López D, Jiménez M, García-Calvo M, Del Val M (2011). Concerted antigen processing of a short viral antigen by human caspase-5 and -10. *J Biol Chem* 286: pp. 16910–3.

[210] Kessler JH, Khan S, Seifert U, Le Gall S, Chow KM, Paschen A, Bres-Vloemans SA, de Ru A, van Montfoort N, Franken KL, Benckhuijsen WE, Brooks JM, van Hall T, Ray K, Mulder A, Doxiadis II, van Swieten PF, Overkleeft HS, Prat A, Tomkinson B, Neefjes J, Kloetzel PM, Rodgers DW, Hersh LB, Drijfhout JW, van Veelen PA, Ossendorp F, Melief CJ (2011). Antigen processing by nardilysin and thimet oligopeptidase generates cytotoxic T cell epitopes. *Nat Immunol* 12: pp. 45–53.

[211] Parmentier N, Stroobant V, Colau D, de Diesbach P, Morel S, Chapiro J, van Endert P, Van den Eynde BJ (2010). Production of an antigenic peptide by insulin-degrading enzyme. *Nat Immunol* 11: pp. 449–54.

[212] Tsuji A, Fujisawa Y, Mino T, Yuasa K (2011). Identification of a plant aminopeptidase with preference for aromatic amino acid residues as a novel member of the prolyl oligopeptidase family of serine proteases. *J Biochem* 150: pp. 525–34.

[213] Rawlings ND, Barrett AJ, Bateman A (2010). *MEROPS*: the peptidase database. *Nucleic Acids Res* 38: pp. D227–33.

[214] Schechter I, Berger A (1967). On the size of the active site in proteases. I. Papain. *Biochem Biophys Res Commun* 27: pp. 157–62.

[215] Timmer JC, Zhu W, Pop C, Regan T, Snipas SJ, Eroshkin AM, Riedl SJ, Salvesen GS (2009). Structural and kinetic determinants of protease substrates. *Nat Struct Mol Biol* 16: pp. 1101–8.

[216] Novotny J, Bruccoleri RE (1987). Correlation among sites of limited proteolysis, enzyme accessibility and segmental mobility. *FEBS Lett* 211: pp. 185–9.

[217] Di Cera E (2009). Serine proteases. *IUBMB Life* 61: pp. 510–5.

[218] Gray GM, Santiago NA (1977). Intestinal surface amino-oligopeptidases. I. Isolation of two weight isomers and their subunits from rat brush border. *J Biol Chem* 252: pp. 4922–8.

[219] Barrett AJ, Rawlings ND (1992). Oligopeptidases, and the emergence of the prolyl oligo-peptidase family. *Biol Chem Hoppe-Seyler* 373: pp. 353–60.

[220] Yang HY, Erdös EG, Levin Y (1971). Characterization of a dipeptide hydrolase (kininase II: angiotensin converting enzyme). *J Pharmacol Exp Ther* 177: pp. 291–300.

[221] Kerr MA, Kenny AJ (1974). The purification and specificity of a neutral endopeptidase from rabbit kidney brush border. *Biochem J* 137: pp. 477–88.

[222] Acharya KR, Sturrock ED, Riordan JF, Ehlers MR (2003). Ace revisited: a new target for structure-based drug design. *Nat Rev Drug Discov* 2: pp. 891–902.

[223] Turner AJ, Isaac RE, Coates D (2001). The neprilysin (NEP). family of zinc metalloendo-peptidases: genomics and function. *BioEssays* 23: pp. 261–9.

[224] Hayashi MA, Camargo ACM (2005). The bradykinin-potentiating peptides from venom

gland and brain of Bothrops jararaca contain highly site-specific inhibitors of the somatic angiotensin-converting enzyme. *Toxicon* 45: pp. 1163–70.

[225] Camargo AC, Graeff FG (1969). Subcellular distribution and properties of the bradykinin inactivation system in rabbit brain homogenates. *Biochem Pharmacol* 18: pp. 548–9.

[226] Camargo AC, Shapanka R, Greene RJ (1973). Preparation, assay, and partial characterization of a neutral endopeptidase from rabbit brain. *Biochemistry* 12: pp. 1838–44.

[227] Oliveira EB, Martins AR, Camargo ACM (1976). Isolation of brain endopeptidases: Influence of size and sequence of substrates structurally related to bradykinin. *Biochemistry* 15: pp. 1967–74.

[228] Carvalho KM, Camargo AC (1981). Purification of rabbit brain endooligopeptidases and preparation of anti-enzyme antibodies. *Biochemistry* 20: pp. 7082–8.

[229] Oliveira ES, Leite PE, Spillantini MG, Camargo AC, Hunt SP (1990). Localization of endo-oligopeptidase (EC 3.4.22.19). in the rat nervous tissue. *J Neurochem* 55: pp. 1114–21.

[230] Medeiros MS, Iazigi N, Camargo ACM, Oliveira EB (1992). Distribution and properties of endo-oligopeptidases A and B in the human endocrine system. *J Endocrinol* 135: pp. 579–88.

[231] Camargo AC, Da Fonseca MJ, Caldo H, De Morais Carvalho K (1982). Influence of the carboxyl terminus of luteinizing hormone-releasing hormone and bradykinin on hydrolysis by brain endo-oligopeptidases. *J Biol Chem* 257: pp. 9265–7.

[232] Greene LJ, Spadaro AC, Martins AR, Perussi De Jesus WD, Camargo AC (1982). Brain endo-oligopeptidase B: a post-proline cleaving enzyme that inactivates angiotensin I and II. *Hypertension* 4: pp. 178–84.

[233] Camargo AC, Almeida ML, Emson PC (1984). Involvement of endo-oligopeptidases A and B in the degradation of neurotensin by rabbit brain. *J Neurochem* 42: pp. 1758–61.

[234] López-Jaramillo P, Antunes-Rodrigues J, Camargo AC (1984). Effect of gonadal steroids on hypothalamus and anterior pituitary endo-oligopeptidase B (proline-endopeptidase). activity in castrated female rats. *Peptides* 5: pp. 1017–9.

[235] Camargo AC, Gomes MD, Toffoletto O, Ribeiro MJ, Ferro ES, Fernandes BL, Suzuki K, Sasaki Y, Juliano L (1994). Structural requirements of bioactive peptides for interaction with endopeptidase 22.19. *Neuropeptides* 26: pp. 281–7.

[236] Camargo AC, Gomes MD, Reichl AP, Ferro ES, Jacchieri S, Hirata IY, Juliano L (1997). Structural features that make oligopeptides susceptible substrates for hydrolysis by recombinant thimet oligopeptidase. *Biochem J* 324: pp. 517–22.

[237] Hayashi MA, Portaro FC, Tambourgi DV, Sucupira M, Yamane T, Fernandes BL, Ferro ES, Rebouças NA, de Camargo AC (2000). Molecular and immunochemical evidences demonstrate that endooligopeptidase A is the predominant cytosolic oligopeptidase of rabbit brain. *Biochem Biophys Res Commun* 269: pp. 7–13.

[238] Gomes MD, Juliano L, Ferro ES, Matsueda R, Camargo AC (1993). Dynorphin-derived peptides reveal the presence of a critical cysteine for the activity of brain endo-oligopeptidase A. *Biochem Biophys Res Commun* 197: pp. 501–7.

[239] Hayashi MA, Portaro FC, Bastos MF, Guerreiro JR, Oliveira V, Gorrão SS, Tambourgi DV, Sant'Anna OA, Whiting PJ, Camargo LM, Konno K, Brandon NJ, Camargo AC (2005). Inhibition of NUDEL (nuclear distribution element-like)-oligopeptidase activity by disrupted-in-schizophrenia 1. *Proc Natl Acad Sci. USA* 102: pp. 3828–33.

[240] Hayashi MA, Portaro FC, Camargo AC (2004). Cytosolic oligopeptidases: features and possible physiopathological roles in the immune and nervous systems. *Curr Med Chem— Central Nervous System Agents* 4: pp. 269–77.

[241] Hayashi MA, Guerreiro JR, Charych E, Kamiya A, Barbosa RL, Machado MF, Campeiro JD, Oliveira V, Sawa A, Camargo AC, Brandon NJ (2010). Assessing the role of endo-oligopeptidase activity of Ndel1 (nuclear-distribution gene E homolog like-1). in neurite outgrowth. *Mol Cell Neurosci* 44: pp. 353–61.

[242] Walter R (1976). Partial purification and characterization of post-proline cleaving enzyme: enzymatic inactivation of neurohypophyseal hormones by kidney preparations of various species. *Biochem Biophys Acta* 422: pp. 138–58.

[243] Moriyama A, Nakanishi M, Sasaki M (1988). Porcine muscle prolyl endopeptidase and its endogenous substrates. *J Biochem* 104: pp. 112–7.

[244] Polgar L (1991). pH-dependent mechanism in the catalysis of prolyl endopeptidase from pig muscle. *Eur J Biochem* 197: pp. 441–7.

[245] Kaushik S, Sowdhamini R (2011). Structural analysis of prolyl oligopeptidases using molecular docking and dynamics: insights into conformational changes and ligand binding. *PLoS ONE* 6: p. e26251.

[246] Orlowski M, Michaud C, Chu TG (1983). A soluble metalloendopeptidase from rat brain. Purification of the enzyme and determination of specificity with synthetic and natural peptides. *Eur J Biochem* 135: pp. 81–8.

[247] Checler F, Vincent JP, Kitabgi P (1983). Degradation of neurotensin by rat brain synaptic membranes: involvement of a thermolysin-like metalloendopeptidase (enkephalinase), angiotensin-converting enzyme, and other unidentified peptidases. *J Neurochem* 41: pp. 375–84.

[248] Rawlings ND, Barrett AJ (1995). Evolutionary families of metallopeptidases. *Methods Enzymol* 248: pp. 183–228.

[249] Checler F, Barelli H, Dauch P, Dive V, Vincent B, Vincent JP (1995). Neurolysin: purification and assays. *Methods Enzymol* 248: pp. 593–614.

[250] Shrimpton CN, Smith AI, Lew RA (2002). Soluble metalloendopeptidases and neuroendocrine signaling. *Endocrine Rev* 23: pp. 647–64.

[251] Fülöp V, Böcskei Z, Polgar L (1998). Prolyl oligopeptidase: an unusual beta-propeller domain regulates proteolysis. *Cell* 94: pp. 161–70.

[252] Brown CK, Madauss K, Lian W, Beck MR, Tolbert WD, Rodgers DW (2001). Structure of neurolysin reveals a deep channel that limits substrate access. *Proc Natl Acad Sci USA* 98: pp. 3127–32.

[253] Natesh R, Schwager SL, Sturrock ED, Acharya KR (2003). Crystal structure of the human angiotensin-converting enzyme-lisinopril complex. *Nature* 421: pp. 551–4.

[254] Johnson GD, Stevenson T, Ahn K (1999). Hydrolysis of peptide hormones by endothelin-converting enzyme-1. A comparison with neprilysin. *J Biol Chem* 274: pp. 4053–8.

[255] Bur D, Dale GE, Oefner C (2001). A three-dimensional model of endothelin-converting enzyme (ECE). based on the X-ray structure of neutral endopeptidase 24.11 (NEP). *Protein Eng* 14: pp. 337–41.

[256] Li M, Chen C, Davies DR, Chiu TK (2010). Induced-fit mechanism for prolyl endopeptidase. *J Biol Chem* 285: pp. 21487–95.

[257] Vincent B, Beaudet A, Dauch P, Vincent JP, Checler F (1996). Distinct properties of neuronal and astrocytic endopeptidase 3.4.24.16: a study on differentiation, subcellular distribution, and secretion processes. *Soc Neurosci* 16: pp. 5049–59.

[258] Palmieri G, Bergamo P, Luini A, Ruvo M, Gogliettino M, Langella E, Saviano M, Hegde RN, Sandomenico A, Rossi M (2011). Acylpeptide hydrolase inhibition as targeted strategy to induce proteasomal down-regulation. *PLoS ONE* 6: p. e25888.

[259] Portaro FC, Gomes MD, Cabrera A, Fernandes BL, Silva CL, Ferro ES, Juliano L, Camargo AC (1999). Thimet oligopeptidase and the stability of MHC class I epitopes in macrophage cytosol. *Biochem Biophys Res Commun* 255: pp. 596–601.

[260] Lev A, Takeda K, Zanker D, Maynard JC, Dimberu P, Waffarn E, Gibbs J, Netzer N, Princiotta MF, Neckers L, Picard D, Nicchitta CV, Chen W, Reiter Y, Bennink JR, Yewdell JW (2008). The exception that reinforces the rule: crosspriming by cytosolic peptides that escape degradation. *Immunity* 28: pp. 787–98.

[261] York IA, Mo AX, Lemerise K, Zeng W, Shen Y, Abraham CR, Saric T, Goldberg AL, Rock KL (2003). The cytosolic endopeptidase, thimet oligopeptidase, destroys antigenic peptides and limits the extent of MHC class I antigen presentation. *Immunity* 18: pp. 429–40.

[262] Saric T, Beninga J, Graef CI, Akopian TN, Rock KL, Goldberg AL (2001). Major histocompatibility complex class I-presented antigenic peptides are degraded in cytosolic extracts primarily by thimet oligopeptidase. *J Biol Chem* 276: pp. 36474–81.

[263] Portaro FC, Hayashi MA, Silva CL, Camargo AC (2001). Free ATP inhibits thimet oligopeptidase (EC 3.4.24.15). activity, induces autophosphorylation in vitro, and controls oligopeptide degradation in macrophage. *Eur J Biochem* 268: pp. 887–94.

[264] Jacchieri SG, Gomes MD, Juliano L, Camargo ACM (1998). A comparative confor-mational analysis of thimet oligopeptidase (EC.3.4.2.4.15). substrates. *J Pept Res* 51: pp. 452–9.

[265] Biedermannova L, Riley EK, Berka K, Hobza P, Vondrasek J (2008). Another role of pro-line: stabilization interactions in proteins and protein complexes concerning proline and tryptophane. *Phys Chem Chem Phys* 10: pp. 6350–9.

[266] Reed J, Reed TA (1997). A set of constructed type spectra for the practical estimation of peptide secondary structure from circular dichroism. *Anal Biochem* 254: pp. 36–40.

[267] Lam TH, Mamitsuka H, Ren EC, Tong JC (2010). TAP Hunter: a SVM-based system for predicting TAP ligands using local description of amino acid sequence. *Immunome Res* 6(Suppl 1): p. S6.

[268] Bjorkman PJ, Saper MA, Samraoui B, Bennett WS, Strominger JL, Wiley DC (1987). Structure of the human class I histocompatibility antigen, HLA-A2. *Nature* 329: pp. 506–12.

[269] Silva CL, Portaro FC, Bonato VL, de Camargo AC, Ferro ES (1999). Thimet oligopep-tidase (EC 3.4.24.15), a novel protein on the route of MHC class I antigen presentation. *Biochem Biophys Res Commun* 255: pp. 591–5.

[270] Maguer-Satta V, Besançon R, Bachelard-Cascales E (2011). Neutral Endopeptidase (CD10): A multifaceted environment actor in stem cells, physiological mechanisms, and cancer. *Stem Cell* 29: pp. 389–96.

[271] Lambert DW, Clarke NE, Turner AJ (2010). Not just angiotensinases: new roles for the angiotensin-converting enzymes. *Cell Mol Life Sci* 67: pp. 89–98.

[272] Mott JD, Werb Z (2004). Regulation of matrix biology by matrix metalloproteinases. *Curr Opin Cell Biol* 16: pp. 558–64.

[273] Sumitomo M, Shen R, Nanus DM (2005). Involvement of neutral endopeptidase in neo-plastic progression. *Biochim Biophys Acta* 1751: pp. 52–9.

[274] Checler F, Vincent JP, Kitabgi P (1986). Purification and characterization of a novel neurotensin-degrading peptidase from rat brain synaptic membranes. *J Biol Chem* 261: pp. 11274–81.

[275] Kato A, Sugiura N, Saruta Y, Hosoiri T, Yasue H, Hirose S (1997). Targeting of endopepti-dase 24.16 to different subcellular compartments by alternative promoter usage. *J Biol Chem* 272: pp. 15313–22.

[276] Wangler NJ, Santos KL, Schadock I, Hagen FK, Escher E, Bader M, Speth RC, Karamyan VT (2012). Identification of membrane-bound variant of metalloendopeptidase neurolysin (EC 3.4.24.16). as the non-angiotensin Type 1 (Non-AT1), Non-AT2 angiotensin binding site. *J Biol Chem* 287: pp. 114–22.

[277]   Brandt I, De Vriendt K, Devreese B, Van Beeumen J, Van Dongen W, Augustyns K, De Meester I, Scharpé S, Lambeir AM (2005). Search for substrates for prolyl oligopeptidase in porcine brain. *Peptides* 26: pp. 2536–46.

[278]   Irazusta J, Larrinaga G, González-Maeso J, Gil J, Meana JJ, Casis L (2002). Distribution of prolyl endopeptidase activities in rat and human brain. *Neurochem Int* 40: pp. 337–45.

[279]   Agirregoitia N, Bizet P, Agirregoitia E, Boutelet I, Peralta L, Vaudry H, Jégou S (2010). Prolyl endopeptidase mRNA expression in the central nervous system during rat development. *J Chem Neuroanat* 40: pp. 53–62.

[280]   Sakaguchi M, Matsuda T, Matsumura E, Yoshimoto T, Takaoka M (2011). Prolyl oligopeptidase participates in cell cycle progression in a human neuroblastoma cell line. *Biochem Biophys Res Commun* 409: pp. 693–8.

[281]   Heng JI, Chariot A, Nguyen L (2009). Molecular layers underlying cytoskeletal remodelling during cortical development. *Trends in Neurosci* 33: pp. 38–47.

[282]   Kisselev AF, Akopian TN, Woo KM, Goldberg AL (1999). The sizes of peptides generated from protein by mammalian 26 and 20 S proteasomes. Implications for understanding the degradative mechanism and antigen presentation. *J Biol Chem* 274: pp. 3363–71.

[283]   Dick LR, Moomaw CR, DeMartino GN, Slaughter CA (1991). Degradation of oxidized insulin B chain by the multiproteinase complex macropain (proteasome). *Biochemistry* 30: pp. 2725–34.

[284]   Lucchiari-Hartz M, van Endert PM, Lauvau G, Maier R, Meyerhans A, Mann D, Eichmann K, Niedermann G (2000). Cytotoxic T lymphocyte epitopes of HIV-1 Nef: Generation of multiple definitive major histocompatibility complex class I ligands by proteasomes. *J Exp Med* 191: pp. 239–52.

[285]   Mo XY, Cascio P, Lemerise K, Goldberg AL, Rock K (1999). Distinct proteolytic processes generate the C and N termini of MHC class I-binding peptides. *J Immunol* 163: pp. 5851–9.

[286]   Fruci D, Niedermann G, Butler RH, van Endert PM (2001). Efficient MHC class I-independent amino-terminal trimming of epitope precursor peptides in the endoplasmic reticulum. *Immunity* 15: pp. 467–76.

[287]   Reits E, Neijssen J, Herberts C, Benckhuijsen W, Janssen L, Drijfhout JW, Neefjes J (2004). A major role for TPPII in trimming proteasomal degradation products for MHC class I antigen presentation. *Immunity* 20(4): pp. 495–506.

[288]   Seifert U, Marañón C, Shmueli A, Desoutter JF, Wesoloski L, Janek K, Henklein P, Diescher S, Andrieu M, de la Salle H, Weinschenk T, Schild H, Laderach D, Galy A, Haas G, Kloetzel PM, Reiss Y, Hosmalin A (2003). An essential role for tripeptidyl peptidase in the generation of an MHC class I epitope. *Nat Immunol* 4: pp. 375–9.

[289] Schatz MM, Peters B, Akkad N, Ullrich N, Martinez AN, Carroll O, Bulik S, Rammensee HG, Endert PV, Tenzer S, Schild H (2008). Characterizing the N-terminal processing motif of MHC class I ligands. *J Immunol* 180: pp. 3210–7.

[290] Princiotta MF, Schubert U, Chen W, Bennink JR, Myung J, Crews CM, Yewdell JW (2001). Cells adapted to the proteasome inhibitor 4-hydroxy- 5-iodo-3-nitrophenylacetyl-Leu-Leu-leucinal-vinyl sulfone require enzymatically active proteasomes for continued survival. *Proc Natl Acad Sci USA* 98: pp. 513–8.

[291] Chang SC, Momburg F, Bhutani N, Goldberg AL (2005). The ER aminopeptidase, ERAP1, trims precursors to lengths of MHC class I peptides by a "molecular ruler" mechanism. *Proc Natl Acad Sci* 102: pp. 17107–12.

[292] Prajapati SC, Chauhan SS (2011). Dipeptidyl peptidase III: a multifaceted oligopeptide N-end cutter. *FEBS* 278: pp. 3256–76.

[293] Mucha A, Drag M, Dalton JP, Kafarski P (2010). Metallo-aminopeptidase inhibitors. *Biochimie* 92: pp. 1509–29.

[294] Gelman JS, Sironi J, Castro LM, Ferro ES, Fricker LD (2011). Peptidomic analysis of human cell lines. *J Proteome Res* 10: pp. 1583–92.

[295] Heemels MT, Ploegh HL (1994). Substrate specificity of allelic variants of the TAP peptide transporter. *Immunity* 1: pp. 775–84.

[296] Evnouchidou I, Momburg F, Papakyriakou A, Chroni A, Leondiadis L, Chang SC, Goldberg AL, Stratikos E (2008). The internal sequence of the peptide-substrate determines its N-terminus trimming by ERAP1. *PLoS ONE* 3: p. e3658.

[297] Peters B, Bui HH, Frankild S, Nielsen M, Lundegaard C, Kostem E, Basch D, Lamberth K, Harndahl M, Fleri W, Wilson SS, Sidney J, Lund O, Buus S, Sette A (2006). A community resource benchmarking predictions of peptide binding to MHC-I molecules. *PLoS Comput Biol* 2(6): p. e65.

[298] Evans DM, et al., (2011). Interaction between ERAP1 and HLA-B27 in ankylosing spondylitis implicates peptide handling in the mechanism for HLA-B27 in disease susceptibility. *Nat Genet* 43: pp. 761–7.

[299] Bachmair A, Finley D, Varshavsky A (1986). In vivo half-life of a protein is a function of its amino-terminal residue. *Science* 234: pp. 179–86.

[300] Hwang CS, Shemorry A, Auerbach D, Varshavasky A (2010). The N-end rule pathway is mediated by a complex of the RING-type Ubr1 and HECT-type UFd4 ubiquitin ligases. *Nat Cell Biol* 12: pp. 1177–85.

[301] Mogk A, Bukau B (2011). When the beginning marks the end. *Science* 327: pp. 966–7.

Author Biographies

Antonio C. M. Camargo received his MD and PhD degrees from the University of São Paulo, Brazil in 1964 and 1969, respectively, having been a graduate student of Mauricio Rocha e Silva, the discoverer of bradykinin. Camargo was a postdoctoral fellow at the Brookhaven National Laboratories, NY (1970–1972), and a visiting researcher at the MRC, Neuropharmacology Unit, New Addenbrooks, Cambridge, UK (1982–1984, Wellcome Foundation Fellowship); at the CNRS at Gif-sur-Yvette, France (1985–1989, INSERM Fellowship); at the University of Kobe-Gakuin, Japan (1990, JICA Fellowship). His academic career was at the University of Sao Paulo (1965–1995). He was the Scientific Director of the Butantan Institute (1997) and founder and director of the Center for Applied Toxinology at the Butantan Institute, Sao Paulo, Brazil (2000–2009). He discovered the first oligopeptidases of the CNS (1979), coining this designation. He has published more than 160 papers on oligopeptidases and bioactive peptides.

Beatriz Lieblich Fernandes received a Biology diploma from the University of Freiburg, Germany, and a PhD in Genetics from Cornell University (1988). She was a Professor of Microbiology at the University of Sao Paulo (1989–2006) and the Coordinator of Institutional Relations and of Intellectual Property Affairs of the Center for Applied Toxinology at the Butantan Institute, Sao Paulo, Brazil (2002–2009).

Lilian Cruz received an undergraduate degree in Biomedicine from the Federal University of Uberlândia, MG, Brazil, in 2011. During this period, she received several young investigator awards for her work on the mechanism of invasion and intracellular traffic of *Trypanosoma cruzi*. In 2011, she entered the Cell Biology program of the University of São Paulo, SP, Brazil, working on the metabolism and function of intracellular peptides.

Emer S. Ferro received a PhD from Paulista Medical School (now UNIFESP) in 1993 and obtained post-doctoral training both at Albert Einstein College of Medicine and Mount Sinai School of Medicine, New York, NY. He has directed a laboratory at the University of São Paulo, Brazil, since 1996. While a graduate student, he discovered that thimet oligopeptidase can be secreted, which is the key to link this enzyme to the metabolism of neuropeptides. In addition, his research group has discovered intracellular peptides and suggested these as a novel class of bioactive molecules within cells. His group also made original contributions to the pharmacology field, with the discovery of hemopressin and its cannabinoid inverse agonist activity. Dr. Ferro has published more than 55 research articles and reviews in the field of peptides and oligopeptidases and is the co-founder of Proteimax Biotechnology.